Current Topics in Microbiology and Immunology

259

Editors

R.W. Compans, Atlanta/Georgia
M. Cooper, Birmingham/Alabama · Y. Ito, Kyoto
H. Koprowski, Philadelphia/Pennsylvania · F. Melchers, Basel
M. Oldstone, La Jolla/California · S. Olsnes, Oslo
M. Potter, Bethesda/Maryland
P.K. Vogt, La Jolla/California · H. Wagner, Munich

Springer
Berlin
Heidelberg
New York
Barcelona
Hong Kong
London
Milan
Paris
Singapore
Tokyo

Nuclear Export of Viral RNAs

Edited by J. Hauber and P.K. Vogt

With 19 Figures and 1 Table

 Springer

Professor Dr. JOACHIM HAUBER
Universität Erlangen-Nürnberg
Institut für Klinische und
Molekulare Virologie
Schlossgarten 4
91054 Erlangen
Germany
e-mail: jmhauber@viro.med.uni-erlangen.de

Professor Dr. PETER K. VOGT
The Scripps Research Institute
Dept. of Molecular and Experimental Medicine
10666 North Torrey Pines Road
La Jolla, CA 92037
USA
e-mail: pkvogt@scripps.edu

Cover Illustration: Nuclear microinjection of HIV-1 Rev in combination with eIF-5A mutant protein reveal eIF-5A-dependent nuclear export of Rev. GST-Rev was microinjected in the nucleus of HeLa cells in combination with GST-eIF-5A proteins and cells were stained with HIV-1 Rev-specific antibodies. Rev is exported to the cytoplasm after 10 min in the presence of eIF-5A wild-type protein (not shown). As shown, coinjection of the an inactive eIF-5A mutant (eIF-5A M13: I135D, T136L) inhibits Rev export (figures courtesy of Dr. Martin Oft, San Francisco, USA)

ISSN 0070-217X
ISBN 3-540-41278-6 Springer-Verlag Berlin Heidelberg New York

This work is subject to copyright. All rights are reserved, whether the whole or part of the material is concerned, specifically the rights of translation, reprinting, reuse of illustrations, recitation, broadcasting, reproduction on microfilm or in any other way, and storage in data banks. Duplication of this publication or parts thereof is permitted only under the provisions of the German Copyright Law of September 9, 1965, in its current version, and permission for use must always be obtained from Springer-Verlag. Violations are liable for prosecution under the German Copyright Law.

Springer-Verlag Berlin Heidelberg New York
a member of BertelsmannSpringer Science + Business Media GmbH

http://www.springer.de

© Springer-Verlag Berlin Heidelberg 2001
Library of Congress Catalog Card Number 15-12910
Printed in Germany

The use of general descriptive names, registered names, trademarks, etc. in this publication does not imply, even in the absence of a specific statement, that such names are exempt from the relevant protective laws and regulations and therefore free for general use.

Product liability: The publishers cannot guarantee the accuracy of any information about dosage and application contained in this book. In every individual case the user must check such information by consulting other relevant literature.

Cover Design: *design & production GmbH*, Heidelberg
Typesetting: Scientific Publishing Services (P) Ltd, Madras
Production Editor: Angélique Gcouta
Printed on acid-free paper   SPIN: 10788723   27/3020 5 4 3 2 1 0

Preface

In eukaryotic cells, the nuclear genome and its transcriptional apparatus is separated from the site of protein synthesis by the nuclear envelope. Thus, a constant flow of proteins and nucleic acids has to cross the nuclear envelope in both directions. This transport in and out of the nucleus is mediated by nuclear pore complexes (NPCs) and occurs in an energy and signal-dependent manner. Thus, nucleocytoplasmic translocation of macromolecules across the nuclear envelope appears to be a highly specific and regulated process. Viruses that replicate their genome in the cell nucleus are therefore forced to develop efficient ways to deal with the intracellular host cell transport machinery. Historically, investigation of Polyomavirus replication allowed identification of sequences that mediate nuclear import, which led subsequently to our detailed understanding of the cellular factors that are involved in nuclear import. Transport of macromolecules in the opposite direction, however, is less well understood. The investigation of retroviral gene expression in recent years provided the first insights into the cellular mechanisms that regulate nuclear export. In particular, the detailed dissection of the function of the human immunodeficiency virus type 1 (HIV-1) Rev *trans*-activator protein identified CRM1, as a *bona fide* nuclear export receptor. CRM1 appears to be involved in the nucleocytoplasmic translocation of the vast majority of viral and cellular proteins that have subsequently been found to contain a Rev-type leucine-rich nuclear export signal (NES). Moreover, increased research in the field of nuclear export of viral mRNAs has also resulted in the description of *cis*-active RNA elements and transport factors in various DNA and RNA viruses that mediate nuclear export by exploiting different and specific cellular nuclear export pathways. In the future it is therefore expected that the continuing investigation of viral RNA export will result in the identification of even more cellular factors that are critically involved in the regulation of nuclear export. Furthermore, it is likely that the elucidation of the interaction of virus RNA transport factors with host cell nuclear export pathways will lead to the identification of novel drug targets to pharmacologically interfere with virus-induced diseases.

We therefore believe that, in the near future, the investigation of viral nuclear RNA export will become an even more exciting area of research. We wish to thank all the contributors and hope that this volume will be of interest to the researchers in the field and will inspire young scientists at the beginning of their research careers.

Erlangen J. HAUBER
La Jolla P.K. VOGT

List of Contents

R.M. Sandri-Goldin
Nuclear Export of Herpes Virus RNA   1

T. Dobner and J. Kzhyshkowska
Nuclear Export of Adenovirus RNA 25

J. Hauber
Nuclear Export Mediated by the Rev/Rex Class
of Retroviral *Trans*-activator Proteins 55

M.-L. Hammarskjöld
Constitutive Transport Element-Mediated
Nuclear Export 77

B. Fahrenkrog, D. Stoffler, and U. Aebi
Nuclear Pore Complex Architecture
and Functional Dynamics 95

R.H. Stauber
Methods and Assays to Investigate Nuclear Export 119

Subject Index 129

List of Contributors

(Their addresses can be found at the beginning of their respective chapters.)

AEBI, U. 95

DOBNER, T. 25

FAHRENKROG, B. 95

HAMMARSKJÖLD, M.-L. 77

HAUBER, J. 55

KZHYSHKOWSKA, J. 25

SANDRI-GOLDIN, R.M. 1

STAUBER, R.H. 119

STOFFLER, D. 95

Nuclear Export of Herpes Virus RNA

R.M. SANDRI-GOLDIN

1	Introduction and General Summary of Herpes Virus RNA Export	1
2	RNA Export of Intron-Containing and Intronless mRNAs During Infection with Herpesviruses	3
2.1	*Cis*-Acting RNA Sequence Elements that Enable Splicing-Independent Expression of the Intronless HSV-1 Thymidine Kinase Gene	3
3	HSV-1 Infection Disrupts Cellular Splicing and Results in the Retention of Unspliced Precursor RNA in the Nucleus	5
4	ICP27 Is Required for the Inhibition of Splicing During HSV-1 Infection and for the Retention of Intron-Containing Transcripts in the Nucleus	6
5	The *Trans*-Acting Viral Protein ICP27 Can Mediate HSV-1 RNA Export by Shuttling Through a Leucine-Rich NES and Binding Intronless RNAs.	8
6	How Does ICP27 Switch from Acting as a Splicing Inhibitor to an RNA Export Protein? . .	11
7	The Regulatory Activities of Epstein-Barr Virus SM Protein, a Homologue of ICP27, and Its Role in RNA Export	14
8	The Regulatory Activities of Homologues of ICP27 in Other Herpesviruses	15
9	Concluding Remarks	16
References		18

1 Introduction and General Summary of Herpes Virus RNA Export

Eukaryotic mRNA precursors are synthesized in the nucleus by RNA polymerase II, after which they are processed by capping at the 5′ end, cleavage and polyadenylation to form the 3′ end, and splicing to remove intron sequences. Following these processing events, the mRNA molecules are transported to the cytoplasm where they are translated. The exchange of macromolecules between the nucleus and cytoplasm occurs through nuclear pore complexes (NPCs). RNA molecules leave the nucleus through an active, energy-dependent, saturable mechanism (for

---

Department of Microbiology and Molecular Genetics, College of Medicine, Medical Sciences I, B240, University of California, Irvine, Irvine, CA 92697-4025, USA

review, see NAKIELNY et al. 1997; MATTAJ and ENGLMEIER 1998; ULLMAN et al. 1997; GORLICH 1998). RNA export through the NPC is mediated by proteins (adapters) associated with the RNAs, and these proteins contain signals to use receptor-mediated export through the NPC to transport their cargo (for review, see STUTZ and ROSBASH 1998). A feature of these proteins is that they shuttle continuously between the nucleus and the cytoplasm, because these export proteins contain a sequence that signals their import into the nucleus, termed a nuclear localization signal (NLS), as well as a sequence that signals their export from the nucleus, termed a nuclear export signal (NES). Regulation of both nuclear import and nuclear export is mainly exerted at the level of transport complex formation. Some of the receptor proteins with which these shuttling export proteins interact have been defined. However, studies on the composition of the NPC, interactions of receptor-cargo complexes with nucleoporin components of the NPC, and the regulation of export and import have only begun to elucidate the transport process.

The study of viruses that utilize and/or modify host cellular pathways can serve as a useful tool in unraveling complex systems. In this regard, the nuclear-replicating herpesviruses, and especially the human α-herpesvirus herpes simplex virus type 1 (HSV-1), is an excellent choice for analyzing nuclear events and cellular trafficking pathways. HSV-1 encodes a large, double-stranded DNA genome comprised of nearly 80 genes that are transcribed in the nucleus by the cellular RNA polymerase II. A curious feature of HSV-1 transcripts is that the majority are intronless and thus, do not undergo pre-mRNA splicing. This presents an unusual situation for the host cell because the best candidate for mRNA export in mammalian cells is the abundant hnRNP A1 protein, which has been shown to shuttle between the nucleus and cytoplasm through a sequence, designated M9, which serves as both the export and import signal (IZAURRALDE et al. 1997). Furthermore, hnRNP A1 was found associated with poly(A)$^+$ RNA in both the nucleus and cytoplasm (PINOL-ROMA and DREYFUSS 1992). It has been postulated that hnRNP A1 binds to the RNA cargo during splicesome assembly (BURD and DREYFUSS 1994) because hnRNP A1 has been shown to play a regulatory role during splicing (CAPUTI et al. 1999; BAI et al. 1999; DEL GATTO-KONCZAK et al. 1999; BLANCHETTE and CHABOT 1999). However, because HSV-1 transcripts do not associate with splicesomes, a separate transport factor is required. Recent studies on herpes virus mRNA export have shown that at least one HSV-1 intronless transcript contains a cis-acting element that interacts with a cellular shuttling protein, hnRNP L (LIU and MERTZ 1995; HUANG et al. 1999b; OTERO and HOPE 1998). It also has been shown that the virus encodes a protein that mediates HSV-1 RNA export. The protein, termed ICP27, contains a NES of the leucine-rich type (SANDRI-GOLDIN 1998a), shuttles between the nucleus and the cytoplasm (SANDRI-GOLDIN 1998a; PHELAN and CLEMENTS 1997; SOLIMAN et al. 1997; MEARS and RICE 1998), and binds RNA in vivo in both the nucleus and the cytoplasm (SANDRI-GOLDIN 1998a). Further, ICP27 preferentially binds to intronless RNAs, rather than spliced mRNAs (SANDRI-GOLDIN 1998a). ICP27 is conserved among the herpesviruses and studies on the homologue in Epstein-Barr virus (EBV) have suggested that it too is involved in RNA export (SEMMES et al. 1998; BOYLE et al. 1999).

This review will summarize what is known about *cis*-acting transport elements in herpes virus mRNAs and the role of the viral regulatory protein ICP27 in RNA export. In addition, the activities and potential role in RNA export of homologues of ICP27 in other herpes viruses, and especially in EBV, will be discussed.

2 RNA Export of Intron-Containing and Intronless mRNAs During Infection with Herpesviruses

2.1 *Cis*-Acting RNA Sequence Elements that Enable Splicing-Independent Expression of the Intronless HSV-1 Thymidine Kinase Gene

Most pre-mRNAs require an intron for efficient processing in higher eukaryotes. Since unspliced messages usually do not leave the nucleus, splicing is generally thought to be intimately associated with mRNA export. In fact, the cellular hnRNP A1 protein, which appears to play a major role in mRNA export (NAKIELNY et al. 1997), also has been shown to function in the regulation of splicing (MAYEDA et al. 1998; BAI et al. 1999; DEL GATTO-KONCZAK et al. 1999; BLANCHETTE and CHABOT 1999). The requirement of a spliceable intron for efficient cytoplasmic accumulation of mRNA has been demonstrated for a number of genes including the SV40 late region (GRUSS et al. 1979), β-globin (HAMER and LEDER 1979; COLLIS et al. 1990), purine nucleoside phosphorylase (JONSSON et al. 1992), mouse thymidylate synthetase (DENG et al. 1989), triosephosphate isomerase (NESIC et al. 1993), and several others. It has been hypothesized that certain splicing factors may interact with nuclear structures, or the spliceosomes may prevent the release and export of RNA until it has been spliced (for review, see NAKIELNY et al. 1997). Retroviruses require viral structural proteins that are encoded by unspliced or partly spliced transcripts; therefore, these viruses have evolved *cis*-acting signals that interact with cellular export factors (KANG and CULLEN 1999; GRUTER et al. 1998; BEAR et al. 1999) and *trans*-acting viral export factors (CULLEN 1992) to bypass the requirement for splicing and to facilitate the transport of unspliced RNA to the cytoplasm.

A number of naturally occurring intronless mRNAs are transported to the cytoplasm without being spliced (HATTORI et al. 1988; HENTSCHEL and BIRNSTIEL 1981; KOILKA et al. 1987; NAGATA et al. 1980). Recently, a number of studies have suggested that efficient export of intronless mRNAs is facilitated by specific sequences within these transcripts. Examples include the hepatitis B virus regulatory element (DONELLO et al. 1998; HUANG and LIANG 1993; HUANG and YEN 1995), and the mouse histone H2a element (HUANG et al. 1999b; HUANG and CARMICHAEL 1997). Among the best studied of the intronless messages is the HSV-1 thymidine kinase (TK) transcript. It has been shown that various non-overlapping regions of the HSV-1 TK gene, when fused upstream or within an intron, can cause the accumulation of cytoplasmic RNA in the absence of intron excision (BUCHMAN and

BERG 1988; GREENSPAN and WEISSMAN 1985; OTERO and HOPE 1998). This suggests that certain *cis*-acting TK sequences function in a positive manner to drive intron-containing transcripts through an export pathway that lacks splicing. LIU and MERTZ (1995) identified an 119-nucleotide sequence element contained within the transcribed region of the TK gene that enabled efficient cytoplasmic accumulation of β-globin RNA in the absence of splicing. They termed this element, which comprises nucleotides 361–479 of the TK transcript, pre-mRNA processing enhancer or PPE. Furthermore, RNA UV-cross-linking assays showed that a 68-kDa protein present in nuclear extracts of HeLa cells specifically bound to the TK PPE element. The 68-kDa protein was found to cross-react with antisera directed against hnRNP L, an abundant 68-kDa cellular protein of previously unknown function. Furthermore, recombinant hnRNP L was shown to bind with high sequence specificity to the TK PPE element, and analysis of substitution mutants in the TK element indicated that hnRNP L binding correlated with the accumulation of RNA in the cytoplasm (LIU and MERTZ 1995). Thus, LIU and MERTZ (1995) concluded that hnRNP L binds in a sequence-specific manner to the TK PPE and enables intron-independent pre-mRNA processing and export. Interestingly, it has recently been shown that hnRNP L interacts with polypyrimidine tract-binding protein (HAHN et al. 1998), which plays a role in splicing. In addition, hnRNP L has also been shown to interact with a unique sequence spanning a 126-nucleotide region in the human and bovine vascular endothelial growth factor (VEGF) 3′-untranslated region, where it is thought to confer mRNA stability under conditions of hypoxia (SHIH and CLAFFEY 1999). The specific binding regions in TK and in VEGF have not been further defined; therefore, it is not clear if hnRNP L is recognizing the same sequence in both transcripts.

In studies to further define the *cis*-acting element in the HSV-1 TK transcript, OTERO and HOPE (1998) substituted regions of the TK gene in a hepatitis B virus (HBV) expression construct to determine if these sequences could functionally replace the HBV posttranscriptional regulatory element by inducing the expression of HBV surface message. Their results supported the conclusion that the TK PPE defined by LIU and MERTZ (1995) was a minimal activity element, and they further identified a total of three *cis*-acting RNA elements in the TK gene. The first region, from nucleotide 60 to 541, included the PPE mapped by LIU and MERTZ (1995), which spans nucleotides 361–479. The two additional elements mapped to TK nucleotides 641–841 and to 941–1141. Similar to the posttranscriptional elements in HBV, duplication of a single TK subelement resulted in greater than additive increases in the cytoplasmic expression of the HBV surface message. These results suggest that intronless genes use a similar strategy for intron-independent expression that requires multiple *cis*-acting sequences. In addition, OTERO and HOPE (1998) found that TK cytoplasmic localization is not sensitive to leptomycin B, a drug that inactivates the CRM-1/exportin pathway (KUDO et al. 1999). Although hnRNP L has not been reported to have a leucine-rich NES, and therefore may use a CRM-1/exportin-independent pathway, it was not determined whether any cellular proteins were bound to the TK sequences; thus it is not clear from these studies whether or not hnRNP L binds to all three TK subelements.

Further studies on the 119 nucleotide HSV-1 TK PPE within a β-globin cDNA vector demonstrated that the viral *cis*-acting element not only appeared to significantly increase the level of cytoplasmic β-globin RNA, but also significantly enhanced the level of polyadenylated globin RNA (HUANG et al. 1999b). The presence of the viral TK element also led to the disappearance of cryptic splicing products (HUANG et al. 1999b). RNA stability analysis indicated that the TK-element-containing, polyadenylated globin mRNAs were not more stable than globin mRNA itself in either the nucleus or the cytoplasm. The TK PPE used in this system functioned like two other *cis*-acting elements from intronless genes, the mouse histone H2a element and the HBV posttranscriptional element. These findings prompted the authors to suggest that a general feature of intronless mRNA transport elements might be a collection of phenotypes, including the inhibition of splicing, the enhancement of polyadenylation, and mRNA export (HUANG et al. 1999b).

3 HSV-1 Infection Disrupts Cellular Splicing and Results in the Retention of Unspliced Precursor RNA in the Nucleus

A feature of the studies that were performed with the HSV-1 TK gene is that in all cases the TK constructs were transfected into cells in the absence of other viral products, or of virus-induced changes in cellular gene expression and nuclear organization. Infection by HSV-1 results in dramatic changes in the nuclear environment in which its gene products are expressed, processed and transported. One such change occurs soon after infection with HSV-1, when the viral *trans*-activator ICP0 migrates to the nuclear domains or ND10, also known as PODs and PML-associated bodies. There it initiates the dispersal of PML and other ND10 antigens (EVERETT and MAUL 1994; MAUL et al. 1996), which have been associated with the control of cellular growth, transcription, and apoptosis (GUIOCHON-MANTEL et al. 1995; MU et al. 1994). Furthermore, HSV-1 DNA replication takes place in large, globular replication compartments (QUINLAN et al. 1984; UPRICHARD and KNIPE 1997), which form at sites adjacent to ND10 (ISHOV and MAUL 1996; LUKONIS and WELLER 1997), and ND10 proteins are recruited to HSV-1 replication sites, suggesting that ND10 play a role in HSV-1 productive infection (BURKHAM et al. 1998).

Another major change that occurs early during viral infection is a dramatic reassortment in the distribution of cellular splicing factors. Monoclonal antibodies directed against components of the small nuclear ribonucleoprotein particles (snRNPs) U1, U2, U4, U5 and U6 or against the essential SR protein splicing factor SC35 show a characteristic speckled or punctate distribution in uninfected cells. HSV-1 infection leads to a coalescence and redistribution in the staining pattern, resulting in the formation of large, globular ball-like structures that move to the periphery of the nucleus as infection proceeds (MARTIN et al. 1987; PHELAN et al. 1993; SANDRI-GOLDIN et al. 1995). Similar changes in the staining pattern of splicing factors have been shown to occur in uninfected cells when splicing was disrupted (O'KEEFE et al. 1994). In fact, infection by HSV-1 appears to impair host cell splicing, resulting in the accumulation of unspliced host and viral pre-mRNAs in the nucleus, and in a decreased accumulation of spliced products in the cytoplasm (SCHRODER et al. 1989; HARDWICKE and SANDRI-GOLDIN 1994). Further, in vitro splicing assays have confirmed that nuclear extracts from HSV-1 infected cells were incompetent to splice an RNA substrate derived from the β-globin gene (HARDY and SANDRI-GOLDIN 1994). Using in situ hybridization labeling methods, PHELAN et al. (1996a) found that intron-containing transcripts from the HSV-1 ICP0 and UL15 genes were increasingly retained in the nucleus in distinct clumps as infection proceeded, and the clumps colocalized with the redistributed snRNPs. In contrast, intronless transcripts were rapidly exported to the cytoplasm (PHELAN et al. 1996a). The genome of HSV-1 encodes nearly 80 transcripts that are expressed during productive infection, and only four of these transcripts contain introns. In addition, because the virus expresses a number of *trans*-activators that stimulate RNA polymerase II transcription of HSV-1 genes, there is a robust expression of HSV-1 transcripts beginning soon after viral entry into the nucleus. Therefore, the inhibition of host cell splicing would not severely affect HSV-1 gene expression and, in fact, would be beneficial to the virus in two ways. First, the disruption of cellular splicing would allow the induction of splicing-independent export pathways so that the abundant viral intronless transcripts would not have to compete with cellular messages in splicing-dependent pathways. Second, cellular unspliced precursor mRNAs would be retained in the nucleus, thus favoring the translation of HSV-1 mRNAs.

4 ICP27 Is Required for the Inhibition of Splicing During HSV-1 Infection and for the Retention of Intron-Containing Transcripts in the Nucleus

The HSV-1 protein that was shown to be responsible for the inhibition of host cell splicing is termed ICP27, also called IE63. ICP27 is a 63-kDa phosphoprotein that is essential for productive viral infection. It is expressed with immediate-early kinetics, and during infection it is required for the switch from early to late gene expression and for maximal levels of DNA synthesis (MCMAHAN and SCHAFFER 1990; UPRICHARD and KNIPE 1996). ICP27 has been shown to function in part at the posttranscriptional level. The first study on its posttranscriptional mechanism of action showed that ICP27 enhanced expression of constructs bearing a synthetic poly(A) signal that lacked a G/U box, and that it reduced expression of constructs containing an intron (SANDRI-GOLDIN and MENDOZA 1992), indicating that ICP27 affected two RNA processing pathways, polyadenylation and splicing. Studies by Clements and colleagues demonstrated that HSV-1 infection appears to alter the specificity of the host polyadenylation machinery in an ICP27-dependent fashion, resulting in a more efficient usage of a subset of viral polyadenylation signals (MCLAUCHLAN et al. 1992; MCGREGOR et al. 1996). UV cross-linking experiments on HeLa nuclear extracts showed that ICP27 enhanced the binding of protein factors, including the 64-kDa component of cleavage stimulation factor (CstF); however the mechanism by which ICP27 increases binding has not been elucidated (MCGREGOR et al. 1996).

The role of ICP27 in the inhibition of splicing was first suggested by evidence that ICP27 mutants are defective in the shut-off of host protein synthesis (HARDWICKE and SANDRI-GOLDIN 1994; SACKS et al. 1985) and that accumulation of spliced RNA from intron-containing target genes was greatly reduced in the presence of ICP27 (SANDRI-GOLDIN and MENDOZA 1992). Subsequent studies showed that the accumulation of unspliced precursor RNA in the nucleus and decreased accumulation of spliced mRNA in the cytoplasm during viral infection required a functional ICP27 protein (HARDWICKE and SANDRI-GOLDIN 1994; PHELAN et al. 1996a). That ICP27 itself was required for the inhibition of splicing was shown using vitro splicing assays utilizing nuclear extracts from cells infected with wild-type HSV-1 or ICP27 mutants (HARDY and SANDRI-GOLDIN 1994). In addition, ICP27 colocalizes with redistributed snRNPs and was found to be both required and sufficient for the relocalization of splicing proteins (PHELAN et al. 1995; SANDRI-GOLDIN et al. 1995). Further, ICP27 co-immunoprecipitates with splicing proteins that react with anti-Sm antisera and with a monoclonal antibody specific for the SR family of splicing proteins, and it appears to alter the phosphorylation state of some of these proteins (SANDRI-GOLDIN and HIBBARD 1996; SANDRI-GOLDIN 1998b). The SR proteins are a highly conserved family of splicing factors that play essential roles in both constitutive and alternative splicing (VALCAREL and GREEN 1996). The interaction of SR proteins with other splicing proteins is regulated by phosphorylation, as is subcellular localization, recruitment to sites of transcription and assembly of spliceosomes (YEAKLEY et al. 1999; MISTELI et al. 1998; WANG et al. 1999; CAO et al. 1997; KOIZUMI et al. 1999). Therefore, we postulate that ICP27 may impair host cell splicing by causing changes in the phosphorylation of some SR proteins (K.S. Sciabica and R.M. Sandri-Goldin, unpublished results). ICP27 has also been shown to alter the phosphorylation of an HSV-1 protein, ICP4, which is the major transcriptional *trans*-activator encoded by the virus (SU and KNIPE 1989; XIA et al. 1996). The mechanism by which ICP27 alters the phosphorylation of proteins is not known. It does not possess recognizable kinase motifs, and it cannot autophosphorylate (ZHI and SANDRI-GOLDIN 1999). It has been shown to interact with SR proteins both in yeast, by two-hybrid analysis, and during viral infection, by co-immunoprecipitation (K.S. Sciabica and R.M. Sandri-Goldin, unpublished results), and it has also been shown to interact with ICP4 (PANAGIOTIDIS et al. 1997). Thus, one possible mechanism might be that interaction with ICP27 occludes phosphorylation sites recognized by cellular kinases.

The region of ICP27 that is required for its effects on splicing and its interaction with splicing factors maps to the C-terminus, from residue 480 to 512, and encompasses a zinc-finger-like domain (Fig. 1) (HARDWICKE et al. 1989; SANDRI-GOLDIN et al. 1995; HARDY and SANDRI-GOLDIN 1994; SANDRI-GOLDIN

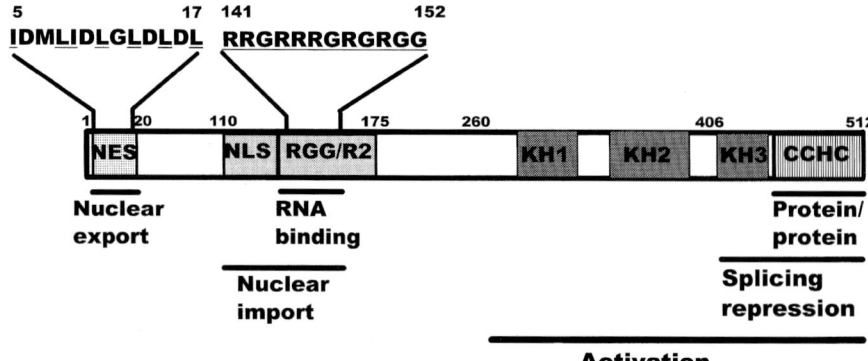

Fig. 1. The 512-amino-acid coding sequence of HSV-1 ICP27 showing the positions of the known functional regions. The sequence of the leucine-rich nuclear export signal (*NES*) from amino acids 5 to 17 is depicted (SANDRI-GOLDIN 1998a), as is the sequence of the RGG box motif, from residues 141 to 152, which is required for RNA binding by ICP27 in vivo and in vitro (SANDRI-GOLDIN 1998a; MEARS and RICE 1996b). The major nuclear localization signal (*NLS*) encompasses amino acids 110–137 (MEARS et al. 1995). Two arginine-rich regions that directly follow the NLS, termed RGG for the RGG motif and R2, are also required for efficient nuclear localization of ICP27 (HIBBARD and SANDRI-GOLDIN 1995). The carboxy-terminal half of the protein, from amino acids 260 to 512, has been shown to be required for the activation of gene expression in transfection assays and during viral infection (HARDWICKE et al. 1989; RICE et al. 1989; RICE and KNIPE 1990; MCMAHAN and SCHAFFER 1990). Three putative KH-like domains, KH1, KH2 and KH3, have recently been identified in the C-terminal half of ICP27 (T.S. Soliman and S.J. Silverstein, personal communication). These domains are postulated to contribute to the RNA binding specificity of ICP27. The region from amino acids 406 to 512 has been shown to be required for the repression of host cell splicing (HARDWICKE et al. 1989; SANDRI-GOLDIN and MENDOZA 1992), and the zinc-finger-like motif within this region must be intact for ICP27 to interact with splicing proteins (SANDRI-GOLDIN et al. 1995; SANDRI-GOLDIN and HIBBARD 1996). The zinc-finger-like domain is also required for ICP27 to self-associate to form multimers in vivo (ZHI et al. 1999). Therefore, this region of the protein appears to be involved in protein-protein interactions

and HIBBARD 1996). In addition, ICP27 interacts with itself in vivo to form multimers and this also requires the zinc-finger region (ZHI et al. 1999). Therefore, the zinc-finger domain appears to be the region involved in at least some of the protein–protein interactions of ICP27.

5 The *Trans*-Acting Viral Protein ICP27 Can Mediate HSV-1 RNA Export by Shuttling Through a Leucine-Rich NES and Binding Intronless RNAs

That ICP27 may shuttle between the nucleus and cytoplasm and thus play a direct role in RNA export apart from its effect on host cell splicing and the subsequent retention of intron-containing RNAs was suggested by the identification of a leucine-rich region in the N-terminus, from amino acids 5 to 17 (Fig. 1). This region exhibits excellent homology to the NESs found in a number of shuttling proteins, including HIV-1 Rev (FISCHER et al. 1995; SZILVAY et al. 1995), PKI (WEN et al.

1995), HTLV-1 Rex (BOGERD et al. 1996; PALMER and MALIM 1996), the Rev proteins of Visna virus and equine infectious anemia virus (MEYER et al. 1996), the E4-34-kDa protein of adenovirus (DOBBELSTEIN et al. 1997), transcription factor III A (FRIDELL et al. 1996), and the cellular hdm2 oncoprotein (ROTH et al. 1998), as well as others. ICP27 was shown to shuttle between the nucleus and cytoplasm in a transcription-dependent manner that can be blocked by the RNA polymerase II inhibitor actinomycin D (SOLIMAN et al. 1997; SANDRI-GOLDIN 1998a; PHELAN and CLEMENTS 1997), and it was also shown to shuttle by heterokaryon analysis (MEARS and RICE 1998). Mutants in which the leucine residues between amino acids 5 and 17 were mutated to arginine or proline residues were unable to move to the cytoplasm, and this was also the case for a mutant in which the NES was deleted (SANDRI-GOLDIN 1998a,b). Further, the fusion of the NES of ICP27 to a green fluorescent protein (GFP) containing a heterologous NLS from the cellular transcription factor LEF (PRIEVE et al. 1996) enabled GFP to shuttle between the nucleus and cytoplasm, confirming that the region from residue 5 to 17 functions as a NES (SANDRI-GOLDIN 1998a). These findings indicate that the leucine-rich NES is both required and sufficient for the shuttling activity (SANDRI-GOLDIN 1998a). However, two studies identified other regions of ICP27 as being involved in the shuttling activity of the protein. A temperature-sensitive mutant with an arginine to histidine substitution at amino acid 480 (SOLIMAN et al. 1997) and a mutant with substitutions at residues 465 and 466 (MEARS and RICE 1998) were unable to shuttle to the cytoplasm. The region of ICP27 between amino acids 260 and 512 has previously been defined as the activator region (Fig. 1), in that this region must be intact for the induction of late gene expression by ICP27 (HARDWICKE et al. 1989; SMITH et al. 1992; RICE and KNIPE 1990). The substitutions in these two mutants do not occur in leucine-rich regions that resemble NES sequences. One explanation for the behavior of these two mutants in shuttling is that overall protein conformation is affected by the mutations at these sites, thereby masking the NES at the amino terminus. An alternate explanation will be discussed below.

Further support that the amino-terminal leucine-rich NES is the major nuclear export signal can be inferred from the finding that insertion of the leucine-rich NES from HIV-1 Rev protein or from the cellular protein PKI into the ICP27 NES deletion mutant restored the ability of ICP27 to shuttle (SANDRI-GOLDIN 1998a,b). These results suggest that, like HIV-1 Rev, ICP27 utilizes the CRM-1/exportin pathway (OSSAREH-NAZARI et al. 1997; FORNEROD et al. 1997; STADE et al. 1997; FUKUDA et al. 1997). Furthermore, export of ICP27 can be blocked by transfecting ten times the amount of a construct that expresses GFP containing the Rev NES (Y. Zhi and R.M. Sandri-Goldin, unpublished results). This suggests that the CRM-1/exportin pathway becomes saturated in the presence of high levels of the Rev NES. The export of hnRNP A1, which uses the transportin pathway (IZAURRALDE et al. 1997), was not greatly affected in these experiments. In addition, SOLIMAN and SILVERSTEIN (2000a) found that export of ICP27 could be blocked by leptomycin B, which inactivates CRM-1-exportin export by a covalent modification at a cysteine residue in the central conserved region of CRM-1 (KUDO et al. 1999). These data strongly support the conclusion that ICP27 is an export protein that shuttles through a leucine-rich NES and uses the CRM-1/exportin pathway.

Another activity that would be required of a protein involved in RNA export is the ability to bind RNA. ICP27 was first shown to bind to single-stranded DNA (VAUGHAN et al. 1992) and RNA in vitro (BROWN et al. 1995; INGRAM et al. 1996c). The region of ICP27 that was both required and sufficient for RNA binding in vitro is an arginine-glycine-rich region (MEARS and RICE 1996b) from amino acids 141 to 152 (Fig. 1) that resembles RGG box motifs found in a number of RNA-binding proteins (KILEDJIAN and DREYFUSS 1992; BIRNEY et al. 1993; GHISOLFI et al. 1992). Furthermore, binding of ICP27 to the 3' region of an intronless, labile interferon mRNA resulted in the accumulation of this RNA in the cytoplasm (BROWN et al. 1995). This accumulation was postulated to result from stabilization of the RNA by ICP27 binding; however, it is also possible to interpret this result as a suggestion that ICP27 is involved in the export of this intronless RNA.

To address whether ICP27 binds RNA during viral infection, UV cross-linking was performed on cells infected with HSV-1. Poly(A)$^+$ RNA was isolated and selected under protein denaturing conditions, such that only proteins in direct contact with RNA would be covalently bound and would co-purify with the poly(A)$^+$ RNA (PINOL-ROMA and DREYFUSS 1992). The RNA-protein complexes were digested with ribonuclease, and the bound proteins were fractionated by polyacrylamide gel electrophoresis. Immunoblot analysis of the bound proteins with monoclonal antibody to ICP27 showed ICP27 bound to both nuclear and cytoplasmic poly(A)$^+$ RNA fractions, further suggesting a direct role in RNA export (SANDRI-GOLDIN 1998a). When the blots were probed with a monoclonal antibody to hnRNP A1, this cellular export protein was recovered in equivalent amounts in nuclear and cytoplasmic RNA fractions from uninfected cells and from cells infected with an ICP27 null-mutant virus. However, the amount of hnRNP A1 that was recovered with poly(A)$^+$ RNA from wild-type HSV-1 infected cells was greatly reduced in both nuclear and cytoplasmic fractions. Western blot analysis of proteins from infected and uninfected cells before UV cross-linking showed that the amount of hnRNP A1 was not altered by virus infection. Thus, the lower amount found bound to RNA was not the result of degradation (SANDRI-GOLDIN 1998a). A possible explanation for the decreased binding of hnRNP A1 to poly(A)$^+$ RNA during HSV-1 infection is that hnRNP A1 may not come into contact with HSV-1 mRNA, because the majority of transcripts are intronless, and because ICP27 impairs splicing and causes the redistribution of splicing factors (HARDY and SANDRI-GOLDIN 1994; SANDRI-GOLDIN et al. 1995; PHELAN et al. 1993). Thus, export pathways used by intron-containing transcripts may be disrupted by virtue of the inhibition of splicing. In contrast, hnRNP L, which has been shown to bind to the PPE of the HSV-1 TK transcript to facilitate its export, was recovered in equivalent amounts in poly(A)$^+$ RNA fractions from uninfected and HSV-1 infected cells. This result supports the hypothesis that hnRNP L is involved in the export of intronless mRNAs.

To determine which RNAs ICP27 binds, in vivo UV cross-linking was performed on HSV-1 infected cells and ICP27 was immunoprecipitated from nuclear and cytoplasmic fractions. Bound RNAs that co-purified with ICP27 were analyzed by RNase protection analysis (SANDRI-GOLDIN 1998a). It was demonstrated that ICP27 bound to seven viral intronless RNAs in both the nucleus and the cytoplasm, and export of these transcripts was diminished substantially during infection with an ICP27 null-mutant virus. This result suggests that ICP27 is required for the export of the transcripts that were analyzed in this study. In contrast, ICP27 did not bind to two HSV-1 mRNAs that contain introns, and both precursor mRNA and spliced forms of these transcripts accumulated in the nucleus during wild-type infection (SANDRI-GOLDIN 1998a). These results strongly suggest that ICP27 plays an important role in the export of HSV-1 intronless mRNAs.

UV cross-linking analysis of ICP27 mutants with deletions or substitutions throughout the coding region showed that the RGG motif from amino acids 141 to 152 was required for RNA binding in vivo (SANDRI-GOLDIN 1998a). This is the region that was shown to be required for RNA binding in vitro (MEARS and RICE 1996b). Furthermore, using a northwestern blotting assay, MEARS and RICE (1996b) showed that the RGG box alone, from residues 140 to 152, can mediate RNA binding when attached to a heterologous protein. This indicates that the RGG motif is not only required, but is also sufficient for binding. Recently, three KH-like RNA-binding motifs (SIOMI et al. 1993; BOMSZTYK et al. 1997; DEJGAARD and LEFFERS 1996) were identified in the C-terminus of ICP27 (SOLIMAN and SILVERSTEIN 2000b). Interestingly, the temperature-sensitive mutant tsLG4, which was found to be defective in shuttling, (SOLIMAN et al. 1997), maps to residue 480 within the putative KH3 region. UV cross-linking studies in cells infected with tsLG4 at the non-permissive temperature showed that tsLG4 ICP27 protein binds RNA much less efficiently than the wild-type protein (Y. Zhi and R.M. Sandri-Goldin, unpublished results). Furthermore, the specificity of binding appears to be affected. Binding to the ICP27 transcript itself was equivalent in tsLG4 and wild-type infected cells, whereas binding to two late transcripts was greatly reduced. These results suggest that these putative KH domains may affect the specificity of RNA binding. Little is known about RNA-binding recognition by ICP27. MEARS and RICE (1996b) found that ICP27 bound efficiently to RNA homopolymers composed of poly(G) and weakly to poly(U) RNA homopolymers. INGRAM et al. (1996c) found a weak RNA-binding activity for ICP27 in their in vitro binding studies, but they were unable to determine any specificity. Thus, it is conceivable that, while the RGG motif is sufficient for RNA binding, the three KH-like regions may determine specificity. The RNA recognition sequences to which ICP27 binds remain to be determined.

6 How Does ICP27 Switch from Acting as a Splicing Inhibitor to an RNA Export Protein?

Studies by SOLIMAN et al. (1997) have shown that shuttling of ICP27 occurs only at late times in infection. Further, in transfected cells, ICP27 shuttling was dependent upon coexpression of RNA from an HSV-1 late gene transcript (SOLIMAN et al. 1997). In addition, we have found that inhibition of cell splicing by ICP27 is most efficient at early times after infection, with splicing activity gradually being restored at late times (HARDY and SANDRI-GOLDIN 1994). These findings suggest that, early in infection, ICP27 is predominantly associated with reassorted splicing complex proteins and is functioning as a splicing inhibitor. As infection progresses and HSV-1 mRNAs become increasingly more abundant, ICP27 begins moving to the cytoplasm, leaving the snRNP-rich nuclear foci where intron-containing RNA accumulates (PHELAN and CLEMENTS 1997) to export intronless RNAs. ICP27 has been shown to bind to immediate-early and early HSV-1 intronless transcripts; however, export of these transcripts occurs in the absence of ICP27, albeit less efficiently (SANDRI-GOLDIN 1998a). In contrast, little cytoplasmic RNA was detected for three HSV-1 late transcripts in infections with an ICP27 null-mutant virus (SANDRI-GOLDIN 1998a). The alterations that are caused by ICP27 to host splicing, the export of intron-containing transcripts and the export of intronless transcripts are summarized diagrammatically in Fig. 2.

At this point, we can only speculate as to what the regulatory switch is. The finding that, in transfected cells, ICP27 shuttling required the presence of an HSV-1 late mRNA implies that ICP27 must bind late RNA to shuttle (SOLIMAN et al. 1997). This notion is further supported by the finding that the tsLG4 protein failed to move to the cytoplasm at the non-permissive temperature (SOLIMAN et al. 1997) and showed a decreased ability to bind to several HSV-1 late mRNAs (Y. Zhi and R.M. Sandri-Goldin, unpublished results). However, in other studies, ICP27 was shown to move to the cytoplasm in transfected cells in the absence of other HSV-1 gene products (SANDRI-GOLDIN 1998a; MEARS and RICE 1998). The possibility that ICP27 also binds cellular intronless transcripts, thus enabling its shuttling activity, cannot be excluded. It should be noted that MEARS and RICE (1996b) found that the RGG box RNA-binding region is methylated in vivo during viral infection. Interestingly, arginine methylation has been shown to facilitate the nuclear export of hnRNP proteins (SHEN et al. 1998; HENRY and SILVER 1996). Recently, KREBBER et al. (1999a) reported that mRNA binding and export could be uncoupled in the yeast export protein Np13p, which is a shuttling hnRNP involved in mRNA processing and transport, and which undergoes methylation (SHEN et al. 1998). Under conditions of stress, Np13p was no longer associated with RNA and was no longer exported through the CRM-1/exportin pathway (KREBBER et al. 1999a). It was postulated that uncoupling Np13p from RNA and the export complex provides a general switch that regulates RNA export and renders Np13p export incompetent and nuclear (KREBBER et al. 1999a). Thus, it is possible that, to turn on the switch, that is, to transform ICP27 from a snRNP-associated splicing inhibitor to a RNA-binding export protein, methylation of the RGG motif is required. To date, no studies have been performed to determine the regulation of methylation of ICP27 or its effect on RNA-binding efficiency. However, a model that has been presented for yeast hnRNP Hrp1p postulates that protein methylation occurs prior to protein-RNA binding, based upon the finding that RNA itself inhibits the methylation of Hrp1p (VALENTINI et al. 1999).

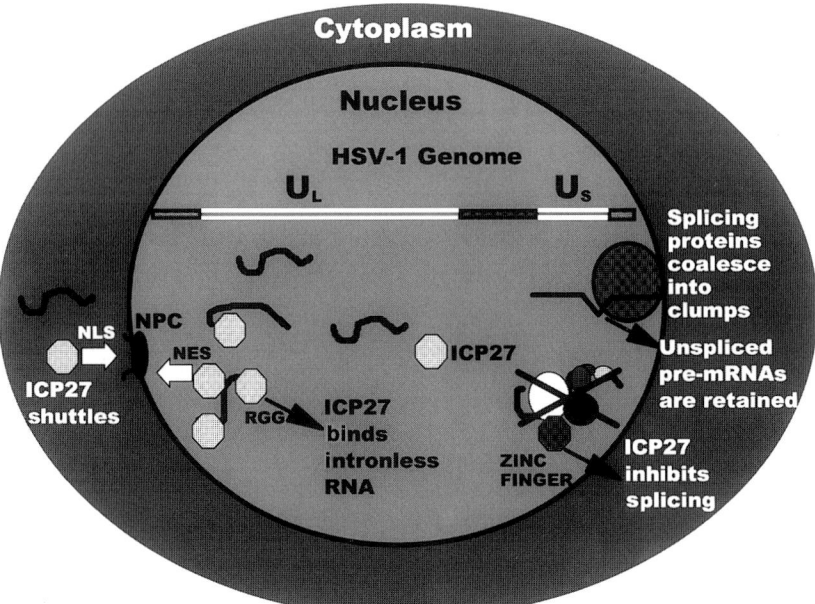

Fig. 2. The role of ICP27 in facilitating the export of HSV-1 intronless mRNAs. Early during infection with HSV-1, ICP27 inhibits host cell splicing and causes the coalescence of snRNPs and other splicing factors into large clumps that move to the nuclear periphery (HARDY and SANDRI-GOLDIN 1994; SANDRI-GOLDIN et al. 1995; MARTIN et al. 1987; PHELAN et al. 1993). The zinc-finger-like domain is required for these activities. Intron-containing pre-mRNAs accumulate in the redistributed clumps (PHELAN et al. 1996a). Later during infection, ICP27 binds HSV-1 intronless mRNAs via the RNA-binding RGG motif (SANDRI-GOLDIN 1998a; MEARS and RICE 1996b) and the protein-RNA complexes move to the cytoplasm through the nuclear pore complex (*NPC*) via a leucine-rich nuclear export signal (*NES*) (SANDRI-GOLDIN 1998a; SOLIMAN et al. 1997; MEARS and RICE 1998; PHELAN and CLEMENTS 1997). Cytoplasmic ICP27 returns to the nucleus via the major nuclear localization signal (*NLS*) (MEARS et al. 1995; HIBBARD and SANDRI-GOLDIN 1995)

Another possible protein modification that might regulate the shuttling activity of ICP27 is protein phosphorylation. ICP27 is phosphorylated during infection and it becomes more highly phosphorylated as infection proceeds (WILCOX et al. 1980; ZHI and SANDRI-GOLDIN 1999). The major phosphorylation sites of ICP27 appear to cluster in the N-terminal portion of the protein, from amino acids 1 to 163 (ZHI and SANDRI-GOLDIN 1999), which encompass both the NES and the NLS. Phosphorylation appears to modulate the efficiency of nuclear import of ICP27, in that a serine to alanine substitution in the protein kinase A (PKA) consensus site, at residue 114 within the NLS (Fig. 1), results in less efficient import of the mutant protein (ZHI and SANDRI-GOLDIN 1999). In addition, we have recently found that deletion of residues 16–18, which encode two serine residues that are phosphorylated by CK II during infection (ZHI and SANDRI-GOLDIN 1999) results in less efficient nuclear export of the mutant ICP27 (Y. Zhi and R.M. Sandri-Goldin, unpublished results). The NES of ICP27 extends from amino acids 5 to 17; thus, the serine at position 16 occurs near the end of the NES, and the serine at position 18 is adjacent to the NES. Accordingly, it is conceivable that the switch that occurs during infection to convert ICP27 into an export protein involves one or more posttranslational modifications of the NES and RNA-binding domains.

7 The Regulatory Activities of Epstein-Barr Virus SM Protein, a Homologue of ICP27, and Its Role in RNA Export

The human γ-herpesvirus EBV encodes a nuclear protein that is a homologue of HSV-1 ICP27. This protein has been termed SM, BS-MLF1, BMLF-1, Mta, and EB2. For simplicity, it will be referred to as SM in this review. The major SM mRNA expressed at early times after entry of EBV into the lytic cycle of replication is a spliced mRNA in which a small upstream exon (BSLF2) is joined to the BMLF1 exon (COOK et al. 1994). SM is one of three EBV-encoded *trans*-activators that regulate EBV lytic cycle gene expression (SEMMES et al. 1998). Like ICP27, SM has been shown to have a posttranscriptional mechanism of action (KENNEY et al. 1989; BUISSON et al. 1989). A role in the enhancement of 3′ processing has been suggested (KEY et al. 1998) because SM enhances the cytoplasmic accumulation of the EBV DNA polymerase transcript, which contains a non-canonical polyadenylation signal. The amount of processed EBV DNA polymerase mRNA was increased several fold, although the rate of transcription was not increased (KEY et al. 1998), suggesting that SM, like ICP27 enhances usage of weak poly(A) sites (KEY et al. 1998). However, another study found no direct enhancement of 3′ end formation by SM in constructs containing either canonical or non-canonical poly(A) sites (BUISSON et al. 1999). Thus, it is not clear whether the observed increase in the accumulation of the EBV DNA polymerase was due to increased processing or to increased RNA export.

SM has also been shown to inhibit the expression of intron-containing genes (RUVOLO et al. 1998) and to associate with splicing factors (SEMMES et al. 1998), suggesting that, like ICP27, SM may affect pre-mRNA splicing. One study presented evidence that SM may not act as a general inhibitor of splicing because SM inhibited the cytoplasmic accumulation of transcripts that were generated by the use of cryptic splice sites or alternative splice sites, but it did not affect the use of constitutive splice sites (BUISSON et al. 1999). Taken together, these findings suggest that SM affects splicing, although this has not been directly tested in vitro in splicing extracts in the presence or absence of SM. The latent and immediate-early genes of EBV contain introns, whereas, many early and late genes expressed during the lytic cycle do not. It has been postulated that SM may regulate EBV lytic infection by down-regulating the synthesis of host cell proteins and latent EBV proteins through effects on splicing, and may simultaneously enhance expression of specific lytic EBV genes by binding mRNA to modulate its stability and export (RUVOLO et al. 1998).

In support of a role in export, SM has been shown to shuttle between the nucleus and cytoplasm (BOYLE et al. 1999; SEMMES et al. 1998). Furthermore, SM is required for the cytoplasmic accumulation of five of the core replication proteins of EBV (SEMMES et al. 1998) as well as several other intronless transcripts (BOYLE et al. 1999). SM has a leucine-rich sequence, from amino acids 226 to 237, with excellent homology to other leucine-rich NES sequences (SEMMES et al. 1998), and mutation of this signal inhibited cytoplasmic translocation and SM activity (BOYLE et al. 1999). Further, movement of SM to the cytoplasm was inhibited by leptomycin B, a specific inhibitor of the CRM-1/exportin pathway, and co-immunoprecipitation experiments showed that SM is associated in vivo with CRM-1 and with other components of the CRM-1 pathway, including the small GTPase Ran and the nucleoporin CAN/Nup214 (BOYLE et al. 1999).

SM has also been shown have RNA-binding activity in that SM-glutathione S-transferase (GST) fusion proteins were found to bind RNA in vitro (RUVOLO et al. 1998; SEMMES et al. 1998) with some specificity, in that two EBV transcripts bound to GST-SM fusion proteins, but a cellular RNA did not bind (SEMMES et al. 1998). The RNA-binding region of SM that is required for its role in export has not been delineated. SM encodes an arginine-X-proline (RXP) tripeptide that is repeated eight times and is similar to the RNA-binding domain in the HSV-1 protein US11 (BUISSON et al. 1999). The RXP repeat region bound RNA as a GST-fusion protein; however, deletion of this region did not abolish the effects of SM on splicing and RNA transport. The C-terminal half of the SM protein is required for these activities (BUISSON et al. 1999). SM is most highly homologous to HSV-1 ICP27 in the C-terminus, and specifically in the zinc-finger-like region, which is required for ICP27's effects on splicing. However, the RGG motif of ICP27 is located in the N-terminal part of the protein, although the KH-like regions map to the C-terminus. It remains to be determined whether SM also possesses KH-like regions and whether they contribute to its ability to bind RNA.

8 The Regulatory Activities of Homologues of ICP27 in Other Herpesviruses

ICP27 is the only HSV-1 immediate early regulatory protein that has homologues not only among other α-herpesviruses, but throughout the herpesvirus family. These homologues include EBV SM protein, discussed above, human varicella-zoster virus (VZV) ORF 4, human cytomegalovirus (CMV) UL69, herpesvirus saimiri (HVS) ORF 57, equine herpesvirus type 1 (EHV-1) UL3, and bovine herpesvirus type 1 (BHV-1) BICP27. The amino acid homology of ICP27 to its counterparts in other herpesviruses is most conserved in residues in the C-terminus (BROWN et al. 1995; PHELAN and CLEMENTS 1998).

Early studies on VZV ORF4, utilizing transfection assays with various target plasmids, demonstrated that, unlike ICP27, VZV ORF4 appeared to act at the transcriptional level, in that activation of the expression of the target genes was dependent upon upstream elements in the promoter regions of these genes (PERERA et al. 1994; MORIUCHI et al. 1994). In addition, ICP27 was unable to complement the activity of ORF4, further suggesting that the proteins functioned differently (MORIUCHI et al. 1994). VZV ORF4 was also found associated with virions and was proposed to play a structural role as a tegument component, similar to the HSV-1 VP16 *trans*-activator (KINCHINGTON et al. 1995). A more recent study found that the activation of two VZV genes by ORF4 tested in transfection assays could occur partly by posttranscriptional means (DEFECHEREUX et al. 1997). Specifically, it was found that increases in CAT mRNA levels did not account for the stimulation in CAT activity that was observed. The posttranscriptional mechanism of ORF4 has not been elucidated, and it is not known if it plays a role in RNA export. Similarly, CMV UL69 protein has been shown to be an *trans*-activator (WINKLER et al. 1994), and like VZV ORF 4, CMV UL69 has been detected in the tegument of virus particles (WINKLER and STAMMINGER 1996). A posttranscriptional mode of action has not been demonstrated for UL69, and it has recently been shown that UL69 arrests cell cycle progression when introduced into uninfected cells by causing cells to accumulate within the G_1 compartment of the cell cycle (LU and SHENK 1999). Similarly, posttranscriptional regulation of gene expression has not been demonstrated for EHV-1 UL3, which has been shown to *trans*-activate an EHV-1 immediate-early protein promoter in CAT assays and to augment the activation, in combination with the immediate-early gene product, of various CAT target genes containing EHV-1 early promoters (ZHAO et al. 1995).

In contrast, evidence has been presented to suggest that two other ICP27 homologues act at the posttranscriptional level. In transfection experiments, it was found that the HVS ORF 57 protein activated or repressed expression from a number of target genes with various HVS promoters. Repression correlated with the presence of an intron in the target gene, and mRNA analysis confirmed that mRNA levels from target genes without introns were not significantly affected, suggesting an effect on splicing (WHITEHOUSE et al. 1998). Furthermore, like ICP27, HVS ORF 57 was found to redistribute the SR protein splicing factor SC35 (COOPER et al. 1999), further suggesting a role in the impairment of splicing during viral lytic infection. BICP27, an early protein expressed by BHV (CHALIFOUR et al. 1996), has been shown to possess a posttranscriptional regulatory function, in that BICP27 up-regulated expression of target genes that differed only in their 3' processing signals (SINGH et al. 1996). Thus, like ICP27, BICP27 may stimulate 3' processing.

9 Concluding Remarks

A striking feature of the *cis*-acting intronless RNA transport elements, and of the viral *trans*-acting export factors encoded by HSV-1 and EBV-1, is that they appear to play multiple roles in RNA processing. HUANG et al. (1999b) suggested that a general feature of intronless RNA elements (and presumably, the cellular proteins which bind to these elements) is a collection of phenotypes, including the inhibition of splicing, the enhancement of polyadenylation, and the facilitation of RNA export. ICP27 has been shown to enhance polyadenylation of selected transcripts (MCLAUCHLAN et al. 1989, 1992; MCGREGOR et al. 1996), and SM has been shown to increase accumulation of a transcript with a weak poly(A) site, although it is not clear if SM can directly enhance 3' processing (BUISSON et al. 1999; KEY et al. 1998). ICP27 has been shown to impair host cell splicing (HARDY and SANDRI-GOLDIN 1994), and SM has also been shown to affect splicing of some intron-containing pre-mRNAs (RUVOLO et al. 1998; BUISSON et al. 1999). In the case of ICP27, the disruption of splicing also results in decreased association of the cellular export protein hnRNP A1 with poly(A)-containing RNA, suggesting that splicing inhibition also interferes with splicing-dependent export pathways. Finally, both ICP27 and SM have been shown to facilitate the export of intronless viral RNAs via a leucine-rich NES through the CRM-1/exportin pathway (SANDRI-GOLDIN 1998a; SEMMES et al. 1998; BOYLE et al. 1999). The export of intronless transcripts that contain *cis*-acting transport elements was found to be independent of the CRM-1/exportin pathway (OTERO and HOPE 1998); however, only one cellular protein that binds to one of these elements has thus far been identified (LIU and MERTZ 1995), and it is likely that there may be other cellular export factors that are involved.

These similarities in phenotypes suggest that, for intronless transcripts to be efficiently exported to the cytoplasm, the nuclear processing, and perhaps the stability of these transcripts, must be increased until cellular or viral factors can bind and facilitate their export through splicing-independent pathways. In the case of viral lytic infection, in which intronless transcripts accumulate rapidly, at least two herpes viruses have evolved proteins that disrupt splicing and splicing-dependent export pathways. It is likely that additional cellular proteins like hnRNP L also function in the export of herpes virus RNAs, and it will be important to identify these proteins. In addition, it will be important to determine whether hnRNP L possesses any activities besides its putative role in export of HSV-1 TK transcripts and its role in stabilizing VEGF transcripts. Does its association with polypyrimidine-tract-binding protein suggest a possible role in splicing inhibition?

The continued study of viral *cis*-acting transport elements and viral export factors and the cellular factors with which they interact will be extremely valuable in unraveling intronless RNA export pathways and their relationship to cellular splicing-dependent pathways.

Acknowledgements. I thank Saul Silverstein (Columbia University) for communicating results prior to publication, and Yan Zhi (University of California, Irvine) for helpful discussions. This work was supported by Public Health Services grant AI21515 from the National Institutes of Allergy and Infectious Diseases.

References

Bai Y, Lee D, Yu T, Chasin LA (1999) Control of 3' splice site choice in vivo by ASF/SF2 and hnRNP A1. Nucleic Acids Res 27:1126–1134

Bear J, Tan W, Zolotukhin AS, Tabernero C, Hudson EA, Felber BK (1999) Identification of novel import and export signals of human TAP, the protein that binds to the constitutive transport element of the type D retrovirus mRNAs. Mol Cell Biol 19:6306–6317

Birney E, Kumar S, Krainer AR (1993) Analysis of the RNA-recognition motif and RS and RGG domains: conservation in metozoan pre-mRNA splicing factors. Nucleic Acids Res 21:5803–5816

Blanchette M, Chabot B (1999) Modulation of exon skipping by high-affinity hnRNP A1-binding sites and by intron elements that repress splice site utilization. EMBO J 18:1939–1952

Bogerd AM, Fridell RA, Benson RE, Hua J, Cullen BR (1996) Protein sequence requirements for function of the human T-cell leukemia virus type 1 Rex nuclear export signal delineated by a novel in vivo randomization-selection assay. Mol Cell Biol 16:4207–4214

Bomsztyk K, Van Seuningen I, Suzuki H, Denisenko O, Ostrowski J (1997) Diverse molecular interactions of the hnRNP K protein. FEBS Lett 403:113–115

Boyle SM, Ruvolo V, Gupta AK, Swaminathan S (1999) Association with the cellular export receptor CRM1 mediates function and intracellular localization of Epstein-Barr virus SM protein, a regulator of gene expression. J Virol 73:6872–6881

Brown CR, Nakamura MS, Mosca JD, Hayward GS, Straus SE, Perera LP (1995) Herpes simplex virus *trans*-regulatory protein ICP27 stabilizes and binds to 3' ends of labile mRNA. J Virol 69:7187–7195

Buchman AR, Berg P (1988) Comparison of intron-dependent and intron-independent gene expression. Mol Cell Biol 8:4395–4405

Buisson M, Hans F, Kusters I, Duran N, Sergeant A (1999) The C-terminal region but not the arg-X-pro repeat of Epstein-Barr virus protein EB2 is required for its effect on RNA splicing and transport. J Virol 73:4090–4100

Buisson M, Manet E, Trescol-Biemont MC, Gruffat H, Durand B, Sergeant A (1989) The Epstein-Barr virus (EBV) early protein EB2 is a posttranscriptional activator expressed under the control of EBV transcription factors EB1 and R. J Virol 63:5276–5284

Burd CG, Dreyfuss G (1994) RNA binding specificity of hnRNP A1: significance of hnRNP A1 high-affinity binding sites in pre-mRNA splicing. EMBO J 13:1197–1204

Burkham J, Coen DM, Weller SK (1998) ND10 protein PML is recruited to herpes simplex virus type 1 prereplicative sites and replication compartments in the presence of viral DNA polymerase. J Virol 72:10100–10107

Cao W, Jamison SF, Garcia-Blanco MA (1997) Both phosphorylation and dephosphorylation of ASF/SF2 are required for pre-mRNA splicing in vitro. RNA 3:1456–1467

Caputi M, Mayeda A, Krainer AR, Zahler AM (1999) hnRNP A/B proteins are required for inhibition of HIV-1 pre-mRNA splicing. EMBO J 18:4060–4067

Chalifour A, Basso J, Gagnon N, Trudel M, Simard C (1996) Transcriptional and translational expression kinetics of the early gene encoding the BICP27 protein of bovine herpesvirus type 1. Virology 224:326–329

Collis P, Antoniou M, Grosveld F (1990) Definition of the minimal requirements within human the β-globin and the dominant control region for high level expression. EMBO J 9:233–240

Cook ID, Shanahan F, Farell PJ (1994) Epstein-Barr virus SM protein. Virology 205:217–227

Cooper M, Goodwin DJ, Hall KT, Stevenson AJ, Meredith DM, Markham AF, Whitehouse A (1999) The gene product encoded by ORF 57 of herpesvirus saimiri regulates the redistribution of the splcinig factor SC-35. J Gen Virol 80:1311–1316

Cullen BR (1992) Mechanism of action of regulatory proteins encoded by complex retroviruses. Microbiol Rev 56:375–394

Defechereux P, Debrus S, Baudoux L, Rentier B, Piette J (1997) Varicella-zoster virus open reading frame 4 encodes an immediate-early protein with posttranscriptional regulatory properties. J Virol 71:7073–7079

Dejgaard K, Leffers H (1996) Characterization of the nucleic-acid-binding activity of KH domains. Different properties of different domains. Eur J Biochem 241:425–431

Del Gatto-Konczak F, Olive M, Gesnel M-C, Breathnach R (1999) hnRNP A1 recruited to an exon in vivo can function as an exon splicing enhancer. Mol Cell Biol 19:251–260

Deng T, Li Y, Johnson LF (1989) Thymidylate synthase gene expression is stimulated by some (but not all) introns. Nucleic Acids Res 17:645–658

Dobbelstein M, Roth J, Kimberly WT, Levine AJ, Shenk T (1997) Nuclear export of the E1B-kDa and E4 34-kDa adenoviral oncoproteins mediated by a rev-like signal sequence. EMBO J 16:4276–4284

Donello JE, Loeb JB, Hope TJ (1998) The woodchuck hepatitis virus contains a tripartite posttranscriptional regulatory element. J Virol 72:5085–5092

Everett RD, Maul GG (1994) HSV-1 IE protein Vmw 110 causes redistribution of PML. EMBO J 13:5062–5069

Fischer U, Huber J, Boelens WC, Mattaj IW, Luhrmann R (1995) The HIV-1 Rev activation domain is a nuclear export signal that accesses an export pathway used by specific cellular RNAs. Cell 82:475–483

Fornerod M, Ohno M, Yoshida M, Mattaj IW (1997) CRM1 is an export receptor for leucine-rich nuclear export signals. Cell 90:1051–1060

Fridell RA, Fischer U, Luhrmann R, Meyer BE, Meinkoth JL, Malim MH, Cullen BR (1996) Amphibian transcription factor IIIA proteins contain a sequence element functionally equivalent to the nuclear export signal of human immunodeficiency virus type 1 Rev. Proc Natl Acad Sci USA 93:2936–2940

Fukuda M, Asano S, Nakamura T, Adachi M, Yoshida M, Yanagida M, Nishida E (1997) CRM1 is responsible for intracellular transport mediated by the nuclear export signal. Nature 390:308–311

Ghisolfi L, Joseph G, Amalric F, Erard M (1992) The glycine-rich domain of nucleolin has an unusual supersecondary structure responsible for its RNA-helix-destabilizing properties. J Biol Chem 267:2955–2959

Gorlich D (1998) Transport into and out of the nucleus. EMBO J 17:2721–2727

Greenspan DS, Weissman SM (1985) Synthesis of predominantly unspliced cytoplasmic RNAs by chimeric herpes simplex virus type 1 thymidine kinase-human b-globin genes. Mol Cell Biol 5:1894–1900

Gruss P, Lai CJ, Dhar R, Khoury G (1979) Splicing as a requirement for biogenesis of functional 16S mRNA of simian virus 40. Proc Natl Acad Sci USA 76:4317–4321

Gruter P, Tabernero C, von Kobbe C, Schmitt C, Saavedra C, Bacchi A, Wilm M, Felber BK, Izaurralde E (1998) TAP, the human homologue of Mex67p, mediates CTE-dependent RNA export from the nucleus. Mol Cell 1:649–659

Guiochon-Mantel A, Savouret JF, Quignon F, Delabre K, Milgrom E, de The H (1995) Effects of PML and PML-ARA on the transactivation properties and subcellular distribution of steroid hormone receptors. Mol Endocrinol 9:1791–1803

Hahn B, Cho OH, Kim JE, Kim YK, Kim JH, Oh YL, Jang SK (1998) Polypyrimidine tract-binding protein interacts with hnRNP L. FEBS Lett 42:401–406

Hamer DH, Leder P (1979) SV40 recombinants carrying a functional RNA splice site junction and polyadenylation site from the chromosomal mouse beta globin gene. Cell 17:737–747

Hardwicke MA, Sandri-Goldin RM (1994) The herpes simplex virus regulatory protein ICP27 can cause a decrease in cellular mRNA levels during infection. J Virol 68:4797–4810

Hardwicke MA, Vaughan PJ, Sekulovich RE, O'Conner R, Sandri-Goldin RM (1989) The regions important for the activator and repressor functions of the HSV-1 alpha protein ICP27 map to the C-terminal half of the molecule. J Virol 63:4590–4602

Hardy WR, Sandri-Goldin RM (1994) Herpes simplex virus inhibits host cell splicing, and regulatory protein ICP27 is required for this effect. J Virol 68:7790–7799

Hattori K, Angel P, Beau MMI, Karin M (1988) Structure and chromosomal localization of the functional intronless human JUN protooncogene. Proc Natl Acad Sci USA 85:9148–9152

Henry MF, Silver PA (1996) A novel methyltransferase (Hmt 1p) modifies poly(A)$^+$-RNA-binding proteins. Mol Cell Biol 16:3668–3678

Hentschel CC, Birnstiel ML (1981) The organization and expression of histone gene families. Cell 25:301–313

Hibbard MK, Sandri-Goldin RM (1995) Arginine-rich regions succeeding the nuclear localization region of the HSV-1 regulatory protein ICP27 are required for efficient nuclear localization and late gene expression. J Virol 69:4656–4667

Huang J, Liang TS (1993) A novel hepatitis B (HBV) genetic element with rev response-like properties that is essential for expression of HBV gene products. Mol Cell Biol 13:7476–7486

Huang Y, Carmichael GG (1997) The mouse histone H2a gene contains a small element that facilitates cytoplasmic accumulation of intronless gene transcripts and of unspliced HIV-1-related mRNAs. Proc Natl Acad Sci USA 94:10104–10109

Huang Y, Wimler KM, Carmichael GG (1999b) Intronless mRNA transport elements may affect multiple steps of pre-mRNA processing. EMBO J 18:1642–1652

Huang Z-M, Yen TSB (1995) Role of the hepatitis B virus posttranscriptional regulatory element in export of intronless transcripts. Mol Cell Biol 15:3864–3869

Ingram A, Phelan A, Dunlop J, Clements JB (1996c) Immediate early protein IE63 of herpes simplex virus type-1 binds RNA directly. J Gen Virol 77:1847–1851

Ishov AM, Maul GG (1996) The periphery of nuclear domain 10 (ND10) as sites of DNA virus deposition. J Cell Biol 134:815–826

Izaurralde E, Jarmolowski A, Beisel C, Mattaj IW, Dreyfuss G (1997) A role for the M9 transport signal of hnRNP A1 in mRNA nuclear export. J Cell Biol 137:27–35

Jonsson JJ, Foresman MD, Wilson N, McIvor RS (1992) Intron requirement for expression of the human purine nucleoside phosphorylase gene. Nucleic Acids Res 20:3191–3198

Kang Y, Cullen BR (1999) The human TAP protein is a nuclear mRNA export factor that contains novel RNA-binding and nucleocytoplasmic transport sequences. Genes Dev 13:1126–1139

Kenney S, Kamine J, Holley-Guthrie E, Mar EC, Lin JC, Markovitz D, Pagano J (1989) The Epstein-Barr virus immediate-early gene product, BMLF1, acts in trans by a posttranscriptional mechanism which is reporter gene dependent. J Virol 63:3870–3877

Key SCS, Yoshizaki T, Pagano JS (1998) The Epstein-Barr virus (EBV) SM protein enhances pre-mRNA processing of the EBV DNA polymerase transcript. J Virol 72:8485–8492

Kiledjian M, Dreyfuss G (1992) Primary structure and binding activity of the hnRNP U protein: binding RNA through RGG box. EMBO J 11:2655–2664

Kinchington PR, Bookey D, Turse SE (1995) The transcriptional regulatory proteins encoded by varicella-zoster virus open reading frames (ORFs) 4 and 63, but not ORF 61, are associated with purified virus particles. J Virol 69:4274–4282

Koilka BK, Frielle T, Collins S, Yang-Feng T, Kobilka TS, Francke U, Lefkowitz RJ, Caron MG (1987) An intronless gene encoding a potential member of the family of receptors coupled to guanine nucleotide regulatory proteins. Nature 329:75–79

Koizumi J, Okamoto Y, Onogi H, Mayeda A, Krainer AR, Hagiwara M (1999) The subcellular localization of SF2/ASF is regulated by direct interaction with SR protein kinases (SRPKs). J Biol Chem 274:11125–11131

Krebber H, Taura T, Lee MS, Silver PA (1999a) Uncoupling of the hnRNP Npl3p from mRNAs during the stress-induced block in mRNA export. Genes Dev 13:1994–2004

Kudo N, Matsumori N, Taoka H, Fujiwara D, Schreiner EP, Wolff B, Yoshida M, Horinouchi S (1999) Leptomycin B inactivates CRM1/exportin 1 by covalent modification at a cysteine residue in the central conserved region. Proc Natl Acad Sci USA 96:9112–9117

Liu X, Mertz JE (1995) HnRNP L binds a *cis*-acting RNA sequence element that enables intron-independent gene expression. Genes Dev 9:1766–1780

Lu M, Shenk T (1999) Human cytomegalovirus UL69 protein induces cells to accumulate in G_1 phase of the cell cycle. J Virol 73:676–683

Lukonis CJ, Weller SK (1997) Formation of herpes simplex virus type 1 replication compartments by transfection: requirements and localization to nuclear domain 10. J Virol 71:2390–2399

Martin TE, Barghusen SC, Leaser GP, Spear PG (1987) Redistribution of nuclear ribonucleoprotein antigens during herpes simplex virus infection. J Cell Biol 105:2069–2082

Mattaj IW, Englmeier L (1998) Nucleocytoplasmic transport: the soluble phase. Annu Rev Biochem 67:265–306

Maul GG, Guidner HH, Everett RD (1996) Nuclear domain 10 as preexisting potential replication start sites of herpes simplex virus type 1. Virology 217:67–75

Mayeda A, Munroe SH, Xu RM, Krainer AR (1998) Distinct functions of the closely related tandem RNA-recognition motifs of hnRNP A1. RNA 4:1111—1123

McGregor F, Phelan A, Dunlop J, Clements JB (1996) Regulation of herpes simplex virus poly(A) site usage and the action of immediate-early protein IE63 in the early-late switch. J Virol 70:1931–1940

McLauchlan J, Phelan A, Loney C, Sandri-Goldin RM, Clements JB (1992) Herpes simplex virus IE63 acts at the posttranscriptional level to stimulate viral mRNA 3′ processing. J Virol 66:6939–6945

McLauchlan J, Simpson S, Clements JB (1989) Herpes simplex virus induces a processing factor that stimulates poly(A) site usage. Cell 59:1093–1105

McMahan L, Schaffer PA (1990) The repressing and enhancing functions of the herpes simplex virus regulatory protein ICP27 map to the C-terminal regions and are required to modulate viral gene expression very early in infection. J Virol 64:3471–3485

Mears WE, Lam V, Rice SA (1995) Identification of nuclear and nucleolar localization signals in the herpes simplex virus regulatory protein ICP27. J Virol 69:935–947

Mears WE, Rice SA (1996b) The RGG box motif of the herpes simplex virus ICP27 protein mediates an RNA-binding activity and determines in vivo methylation. J Virol 70:7445–7453

Mears WE, Rice SA (1998) The herpes simplex virus immediate-early protein ICP27 shuttles between the nucleus and cytoplasm. Virology 242:128–137

Meyer BE, Meinkoth JL, Malim MH (1996) Nuclear transport of human immunodeficiency virus type 1, visna virus, and equine infectious anemia virus rev proteins: identification of a family of nuclear export signals. J Virol 70:2350–2359

Misteli T, Caceres JF, Clement JQ, Krainer AR, Wilkinson MF, Spector DL (1998) Serine phosphorylation of SR proteins is required for their recruitment to sites of transcription in vivo. J Cell Biol 143:297–307

Moriuchi H, Moriuchi M, Smith HA, Cohen JI (1994) Varicella-zoster virus open reading frame 4 protein is functionally distinct from and does not complement its herpes simplex virus type 1 homolog ICP27. J Virol 68:1987–1992

Mu Z, Chin KV, Liu JH, Lozano G, Chang KS (1994) PML, a growth suppressor disrupted in acute promyelocytic leukemia. Mol Cell Biol 14:6858–6867

Nagata S, Mantei N, Weissman C (1980) The structure of one of the eight or more distinct chromosomal genes for human interferon-a. Nature 287:401–408

Nakielny S, Fischer U, Michael WM, Dreyfuss G (1997) RNA transport. Annu Rev Neurosci 20:269–301

Nesic D, Cheng J, Maquat LE (1993) Sequences within the last intron function in RNA 3′-end formation in cultured cells. Mol Cell Biol 13:3359–3369

O'Keefe RT, Mayeda A, Sadowski CL, Krainer AR, Spector DL (1994) Disruption of pre-mRNA splicing in vivo results in reorganization of splicing factors. J Cell Biol 124:249–260

Ossareh-Nazari B, Bachelerie F, Dargemont C (1997) Evidence for a role of CRM1 in signal-mediated nuclear protein export. Science 278:141–144

Otero GC, Hope TJ (1998) Splicing-independent expression of the herpes simplex virus type 1 thymidine kinase gene is mediated by three *cis*-acting RNA subelements. J Virol 72:9889–9896

Palmer D, Malim MH (1996) The human T-cell leukemia virus type 1 posttranscriptional *trans*-activator Rex contains a nuclear export signal. J Virol 70:6442–6445

Panagiotidis CA, Lium EK, Silverstein S (1997) Physical and functional interactions between herpes simplex virus immediate-early proteins ICP4 and ICP27. J Virol 71:1547–1557

Perera LP, Kaushal S, Kinchington PR, Mosca JD, Hayward GS, Straus SE (1994) Varicella-zoster virus open reading frame 4 encodes a transcriptional activator that is functionally distinct from that of herpes simplex virus homology ICP27. J Virol 68:2468–2477

Phelan A, Carmo-Fonseca M, McLauchlan J, Lamond AI, Clements JB (1993) A herpes simplex virus type 1 immediate-early gene product, IE63, regulates small nuclear ribonucleoprotein distribution. Proc Natl Acad Sci USA 90:9056–9060

Phelan A, Clements JB (1997) Herpes simplex virus type 1 immediate early protein IE63 shuttles between nuclear compartments and the cytoplasm. J Gen Virol 78:3327–3331

Phelan A, Clements JB (1998) Posttranscriptional regulation in herpes simplex virus. Sem Virol 8: 309–318

Phelan A, Dunlop J, Clements JB (1996a) Herpes simplex virus type 1 protein IE63 affects the nuclear export of virus intron-containing transcripts. J Virol 70:5255–5265

Pinol-Roma S, Dreyfuss G (1992) Shuttling of pre-mRNA binding proteins between nucleus and cytoplasm. Nature 355:730–732

Prieve MG, Guttridge KL, Mungia JE, Waterman ML (1996) The nuclear localization signal of lymphoid enhancer factor-1 is recognized by two differentially expressed Srp1-nuclear localization sequence receptor proteins. J Biol Chem 271:7654–7658

Quinlan MP, Chen LB, Knipe DM (1984) The intranuclear location of a herpes simplex virus DNA-binding protein is determined by the status of viral DNA replication. Cell 36:657–668

Rice SA, Knipe DM (1990) Genetic evidence for two distinct transactivation functions of the herpes simplex virus alpha protein ICP27. J Virol 64:1704–1715

Rice SA, Su L, Knipe DM (1989) Herpes simplex virus alpha protein ICP27 possesses separable positive and negative regulatory activities. J Virol 63:3399–3407

Roth J, Dobbelstein M, Freedman DA, Shenk T, Levine AJ (1998) Nucleo-cytoplasmic shuttling of the hdm2 oncoprotein regulates the levels of the p53 protein via a pathway used by the human immunodeficiency virus rev protein. EMBO J 17:554–564

Ruvolo V, Wang E, Boyle S, Swaminathan S (1998) The Epstein-Barr virus nuclear protein SM is both a post-transcriptional inhibitor and activator of gene expression. Proc Natl Acad Sci USA 95:8852–8857

Sacks WR, Greene CC, Ashman DP, Schaffer PA (1985) Herpes simplex virus type 1 ICP27 is an essential regulatory protein. J Virol 55:796–805

Sandri-Goldin RM (1998a) ICP27 mediates herpes simplex virus RNA export by shuttling through a leucine-rich nuclear export signal and binding viral intronless RNAs through an RGG motif. Genes Dev 12:868–879

Sandri-Goldin RM (1998b) Interactions between an HSV regulatory protein and cellular mRNA processing pathways. Methods 16:95–104

Sandri-Goldin RM, Hibbard MK (1996) The herpes simplex virus type 1 regulatory protein ICP27 coimmunoprecipitates with anti-Sm antiserum and an the C-terminus appears to be required for this interaction. J Virol 70:108–118

Sandri-Goldin RM, Hibbard MK, Hardwicke MA (1995) The C-terminal repressor region of HSV-1 ICP27 is required for the redistribution of small nuclear ribonucleoprotein particles and splicing factor SC35; however, these alterations are not sufficient to inhibit host cell splicing. J Virol 69:6063–6076

Sandri-Goldin RM, Mendoza GE (1992) A herpes virus regulatory protein appears to act posttranscriptionally by affecting mRNA processing. Genes Dev 6:848–863

Schroder HC, Falke D, Weise K, Bachman M, Carmo-Fonseca M, Zaubitzer T, Muller WEG (1989) Change of processing and nucleocytoplasmic transport on mRNA in HSV-1 infected cells. Virus Res 13:61–78

Semmes GJ, Sarisky RT, Gao Z, Zhong L, Hayward SD (1998) Mta has properties of an RNA export protein and increases cytoplasmic accumulation of Epstein-Barr virus replication gene mRNA. J Virol 72:9526–9534

Shen EC, Henry MF, Weiss VH, Valentini SR, Silver PA, Lee MS (1998) Arginine methylation facilitates the nuclear export of hnRNP proteins. Genes Dev 12:679–691

Shih S-C, Claffey KP (1999) Regulation of human vascular endothelial growth factor mRNA stability in hypoxia by heterogeneous nuclear ribonucleoprotein L. J Biol Chem 274:1365

Singh M, Fraefel C, Bello LJ, Lawrence WC, Schwyzer M (1996) Identification and characterization of BICP27, an early protein of bovine herpesvirus 1 which may stimulate mRNA 3' processing. J Gen Virol 77:615–625

Siomi H, Matunis MJ, Michael WM, Dreyfuss G (1993) The pre-mRNA binding K protein contains a novel evolutionarily conserved motif. Nucleic Acids Res 21:1193–1198

Smith IL, Hardwicke MA, Sandri-Goldin RM (1992) Evidence that the herpes simplex virus immediate early protein ICP27 acts post-transcriptionally during infection to regulate gene expression. Virology 186:74–86

Soliman TM, Sandri-Goldin RM, Silverstein S (1997) Shuttling of the herpes simplex virus type 1 regulatory protein ICP27 between the nucleus and cytoplasm mediates the expression of late proteins. J Virol 71:9188–9197

Soliman TM, Silverstein SJ (2000a) Herpes virus mRNAs are sorted for export via CRM1-dependent and -independent pathways. J Virol 74:2814–2825

Soliman TM, Silverstein SJ (2000b) Identification of an export control sequence and a requirement for the KH domains in ICP27 from herpes simplex virus type 1. J Virol 74:7600–7609

Stade K, Ford CS, Guthrie C, Weis K (1997) Exportin 1 (CRM1p) is an essential nuclear export factor. Cell 90:1041–1050

Stutz F, Rosbash M (1998) Nuclear RNA export. Genes Dev 12:3303–3319

Su L, Knipe DM (1989) Herpes simplex virus alpha protein ICP27 can inhibit or augment viral gene transactivation. Virology 170:496–504

Szilvay AM, Brokstad KA, Kopperrud R, Haukenes G, Kalland K-H (1995) Nuclear export of the human immunodeficiency virus type 1 nucleocytoplasmic shuttle protein Rev is mediated by its activation domain and is blocked by transdominant negative mutants. J Virol 69:3315–3323

Ullman KS, Powers MA, Forbes DJ (1997) Nuclear export receptors: from importin to exportin. Cell 90:967–970

Uprichard SL, Knipe DM (1996) Herpes simplex ICP27 mutant viruses exhibit reduced expression of specific DNA replication genes. J Virol 70:1969–1980

Uprichard SL, Knipe DM (1997) Assembly of herpes simplex virus replication proteins at two distinct intranuclear sites. Virology 229:113–125

Valcarel J, Green MR (1996) The SR protein family: pleiotropic functions in pre-mRNA splicing. Trends Biochem Sci 21:296–301

Valentini SR, Weiss VH, Silver PA (1999) Arginine methylation and binding of Hrp1p to the efficiency element for mRNA 3'-end formation. RNA 5:272–280

Vaughan PJ, Thibault KJ, Hardwicke MA, Sandri-Goldin RM (1992) The herpes simplex virus type 1 immediate early protein ICP27 encodes a potential metal binding domain and is able to bind to zinc. Virology 189:377–384

Wang X, Bruderer S, Rafi Z, Xue J, Milburn PJ, Kramer A, Robinson PJ (1999) Phosphorylation of splicing factor SF1 on ser20 by cGMP-dependent protein kinase regulates splicesome assembly. EMBO J 18:4549–4559

Wen W, Meinkoth JL, Tsien RY, Taylor SS (1995) Identification of a signal for rapid export of proteins from the nucleus. Cell 82:463–473

Whitehouse A, Cooper M, Meredith DM (1998) The immediate-early gene product encoded by open reading frame 57 of herpesvirus saimiri modulates gene expression at a posttranscriptional level. J Virol 72:857–861

Wilcox KW, Kohn A, Sklyanskaya E, Roizman B (1980) Herpes simplex virus phosphoproteins. I. Phosphate cycles on and off some viral polypeptides and can alter their affinity for DNA. J Virol 33:167–182

Winkler M, Rice SA, Stamminger T (1994) UL69 of human cytomegalovirus, an open reading frame with homology to ICP27 of herpes simplex virus, encodes a transactivator of gene expression. J Virol 68:3943–3954

Winkler M, Stamminger T (1996) A specific subform of the human cytomegalovirus transactivator protein pUL69 is contained within the tegument of virus particles. J Virol 70:8984–8987

Xia K, DeLuca NA, Knipe DM (1996) Analysis of phosphorylation sites of herpes simplex virus type 1 ICP4. J Virol 70:1061–1071

Yeakley JM, Tronchere H, Olesen J, Dyck JA, Wang H-Y, Fu XD (1999) Phosphorylation regulates in vivo interaction and molecular targeting of serine/arginine-rich pre-mRNA splicing factors. J Cell Biol 145:447–455

Zhao Y, Holden VR, Smith RH, O'Callaghan DJ (1995) Regulatory function of the equine herpesvirus 1 ICP27 gene product. J Virol 69:2786–2793

Zhi Y, Sandri-Goldin RM (1999) Analysis of the phosphorylation sites of the herpes simplex virus type 1 regulatory protein ICP27. J Virol 73:3246–3257

Zhi Y, Sciabica C, Sandri-Goldin RM (1999) Self interaction of the herpes simplex virus type 1 regulatory protein ICP27. Virology 257:341–351

Nuclear Export of Adenovirus RNA

T. Dobner and J. Kzhyshkowska

1	Introduction	25
2	Adenovirus Infectious Cycle	27
3	Nuclear Export of Adenovirus RNA	29
3.1	RNA Export: General Considerations	29
3.2	Evidence for Regulated RNA Export in Adenovirus-Infected Cells	31
3.3	Regulated Viral RNA Export Is Linked to Alterations in Nuclear Organization	32
3.4	E4orf3 and E4orf6 Maintain the Stability of Major Late RNAs in the Nucleus	34
3.5	E1B-55kDa/E4orf6 Counteract Nuclear Retention of Viral RNAs	36
3.6	E1B-55kDa and E4orf6 Are Viral Shuttle Proteins	39
3.7	Requirement for E1B-55kDa and E4orf6 in Late Viral RNA Export Is Dependent on the Type of Host Cell	43
4	Involvement of E1B-55kDa in Translational Control	44
5	Involvement of E1B-55kDa and E4orf6 in Posttranslational Control	46
6	Conclusions and Future Directions	47
	References	49

1 Introduction

Adenoviruses are medium-sized, non-enveloped DNA viruses which nowadays include well over 100 different serotypes found in a wide range of mammalian and avian hosts. The human adenoviruses comprise over 47 different serotypes which cause lytic and persistent infections and have been associated with a variety of clinical syndromes (reviewed in Horwitz 1996). Following their discovery in latently infected adenoids more than 45 years ago, human adenoviruses stepped into the limelight of molecular virology when it was found that certain serotypes have oncogenic potential in newborn rodents and that all human adenoviruses can transform primary rodent cells in culture. These findings in particular inspired a period of intense research on adenovirus biology that contributed enormously to

---

Institut für Medizinische Mikrobiologie und Hygiene, Universität Regensburg, Franz-Josef-Strauß-Allee 11, 93053 Regensburg, Germany

a molecular understanding of normal and malignant cell growth. In addition, studies on adenovirus productive infection in cultured cells have provided important insight into fundamental mechanisms in molecular biology, perhaps most notably, mRNA splicing. While these viruses still serve as a laboratory model to solve the mysteries of cell growth control, they are now being used in pre-clinical and clinical trials as vectors for gene therapy and more recently as oncolytic vehicles for the treatment of human cancer (reviewed in BILBAO et al. 1998; BENIHOUD et al. 1999).

Human adenoviruses have small, linear, double-stranded DNA genomes encoding only the virus structural polypeptides and a limited number of regulatory proteins. In order to be able to maintain a compact genome size, and hence restricted coding capacity, yet remain an effective pathogen, these viruses have evolved efficient strategies to utilize the host cell machinery for viral gene expression, DNA replication and formation of progeny virions. Their ability to subvert host macromolecular synthesis at defined stages of the viral replication cycle is the result of both the viral program of transcriptional regulation and virus-encoded functions that control mRNA processing, nucleocytoplasmic mRNA transport and translation. It is the posttranscriptional mechanisms in particular which serve to facilitate production of large quantities of viral structural polypeptides and to shut off host cell protein synthesis at late times of infection. In productively infected cells, regulated viral mRNA splicing and viral mRNA export are controlled by at least three adenovirus early proteins, E4orf3 (early region 4, open reading frame 3), E4orf6 and E1B-55kDa (early 1B 55-kDa protein). These directly or indirectly alter the pattern of splicing and stability of late viral RNAs and induce changes in the export of cellular and viral transcripts, resulting in the selective cytoplasmic accumulation of viral mRNAs. It appears that the functional moiety involved in viral mRNA export is a physical complex between the E1B-55kDa and E4orf6 proteins. Several observations indicate that these E1B-55kDa/E4orf6 complexes mediate viral mRNA export through mechanisms that block nuclear retention of late mRNAs retaining characteristics of unspliced transcripts. More recently, it has been proposed that E1B-55kDa/E4orf6 directly participate in nucleocytoplasmic RNA trafficking, since both proteins possess leucine-rich nuclear export signals and shuttle between the nuclear and the cytoplasmic compartments. Altogether, these data suggest substantial mechanistic similarities between adenovirus and complex retroviruses in the control of viral RNA export. This chapter summarizes the available information on nuclear export of adenovirus RNA with particular emphasis on current developments. We describe several important aspects of regulated mRNA transport in adenovirus-infected cells and discuss models that may account for reciprocal effects on viral and cellular mRNA export. Additional information on virally regulated mRNA processing and translational control of adenovirus gene expression may be found in several excellent reviews of the subject (ZHANG and SCHNEIDER 1993; IMPERIALE et al. 1995; LEPPARD 1997, 1998).

2 Adenovirus Infectious Cycle

The vast majority of the work on the adenovirus replicative cycle in cultured cells has focused on the closely related group C adenoviruses types 2 and 5 (Ad2 and Ad5; reviewed in SHENK 1996). These serotypes are easily grown in tissue culture and their viral genome is readily manipulated, facilitating the study of viral gene functions by mutational analyses. Studies with other serotypes indicate that human adenoviruses have the same genomic organization, express a similar set of RNAs and regulate their expression by equivalent processes. Ad2 and Ad5 have characteristic linear, double-stranded DNA genomes approximately 36kb in length (Fig. 1). Their viral chromosome carries eight RNA-polymerase-II-dependent transcription units, which encode around 40 different polypeptides. In addition, two segments on their viral genome encode for small RNAs (VA RNAs I and II) of about 160 nucleotides which are transcribed by RNA polymerase III.

As for most DNA viruses, the infectious cycle of Ad2 and Ad5 is temporally organized into early and late phases, which are separated by the onset of viral DNA replication. The early phase of infection is characterized by the production of viral mRNAs from the four early transcription units, designated E1–E4, and from late region 1 (L1), which is transcribed early from the major late promoter (MLP). These transcription units produce multiple differentially spliced and polyadenylated mRNAs encoding a variety of distinct polypeptides. Early viral gene products function in transcriptional and posttranscriptional regulation of viral and host RNA production, induce cell cycle progression and antagonize or suppress a variety of anti-viral defense mechanisms, such as apoptosis and the immune response of the host organism. Expression of early genes is required to establish an optimal environment for viral DNA replication, which in turn is necessary for expression of viral late genes. In general, the early phase of the infectious cycle is not associated with dramatic changes in cellular protein synthesis. Following the onset of viral DNA replication, adenovirus late transcription is activated at three different transcription units, which mostly encode for structural polypeptides or are involved in packaging of viral genomic DNA (Fig. 1). With the exception of IVa2 and IX, the adenovirus late coding regions are organized into a single large transcription unit, called the major late transcription unit (MLTU). The MLTU is controlled by the MLP and generates a large primary transcript of approximately 29,000 nucleotides in length that is processed by differential polyadenylation and splicing into at least 18 cytoplasmic mRNAs. Based on the utilization of common poly (A) addition sites, these mRNAs have been grouped into five families (L1–L5). All mRNAs transcribed from the MLP contain a common 5′ non-coding region of 201 nucleotides termed the tripartite leader (TPL), derived by the splicing of three small exons. Within a few hours of the onset of the late phase, several complex metabolic changes occur that ensure preferential synthesis of viral proteins and efficient assembly of progeny virions. Late mRNAs are preferentially exported to the cytoplasm where they are translated to the exclusion of host mRNAs. In addition, cytoplasmic accumulation of most cellular mRNAs is blocked. As a result, very

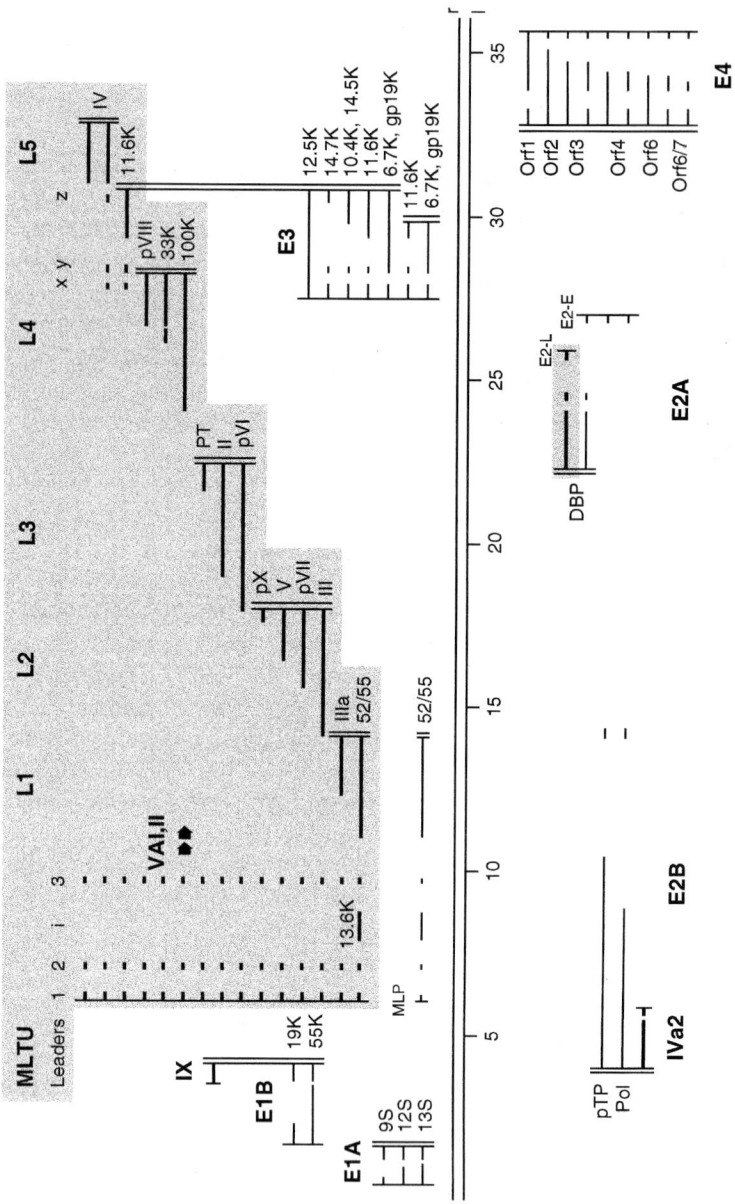

Fig. 1. Map of genes and transcription units encoded by the adenovirus type 2/5 genomes. The linear double-stranded genome is depicted in the *center* of the figure, with *r* and *l* referring to *rightward* (*above*) and *leftward* transcription (*below*). Transcription units are named in *boldface*. *Thin lines* indicate early mRNAs and *thick lines* indicate mRNAs (IX, IVa2 and E2B) expressed at intermediate times of infection. *Thick lines* within the *gray shaded area* (MLTU and E2A) show mRNAs expressed late after infection. Individual mRNA species are shown as *solid lines*. *Gaps in the lines* denote the position of introns. RNA polymerase II-dependent promoters are shown as *solid vertical lines* and common polyadenylation sites as *paired vertical lines*. *Bold black arrows* indicate virus-associated RNAs (VA RNA I and II) transcribed by RNA polymerase III. Selected polypeptides that have been assigned to different regions are indicated adjacent to the RNA sequence encoding them. *MLTU*, major late transcription unit; *ORF*, open reading frame; *PT*, 23K virion proteinase; *gp* glycoprotein; *DBP*, DNA-binding protein; *pTP*, pre-terminal protein; *Pol*, DNA polymerase; *MLP*, major late promoter; *E2-E, E2-L*, E2 promoters used at early and late times of infection. For the sake of clarity several RNA species are not included if the predicted product is the same or similar to one shown, or where no unique function for the RNA/protein has as yet been identified. (Adapted from Imperiale et al. 1995 and Leppard 1998)

large quantities of viral structural polypeptides are produced while cellular protein synthesis is shut off almost completely. These profound alterations in viral and host macromolecular synthesis involve several virus-encoded regulatory components that operate at the level of nucleocytoplasmic mRNA transport and translation. The first process is the major subject of this review. In HeLa cells infected with Ad2 or Ad5 at high multiplicities (10–200 plaque-forming units per cell), the lytic cycle is completed approximately 24h after infection. At the end of the cycle, the infected cell has increased its total DNA and protein content two-fold and approximately 10,000 virus particles per infected cell have been produced.

3 Nuclear Export of Adenovirus RNA

3.1 RNA Export: General Considerations

Over the past 10 years, considerable progress has been made in identifying molecular mechanisms underlying the active nuclear export of RNA (reviewed in Mattaj and Englmeier 1998; Nakielny and Dreyfuss 1999). It is generally now well accepted that different classes of RNAs (e.g. mRNA, tRNA, rRNA, and snRNA) exit the nucleus via distinct pathways in the form of ribonucleoprotein complexes (RNPs). A large body of evidence suggests that RNA export through the nuclear pore complex (NPC) is a receptor-mediated process that requires the concerted interaction between soluble factors (export adaptors and export receptors) that shuttle between the nuclear and the cytoplasmic compartment and nuclear pore components, as well as mechanisms that regulate retention and release. To date the best characterized RNA transport factor is the human immunodeficiency virus type 1 (HIV-1) Rev protein. In fact much of what we know today about the mechanics of RNA export originates from studying the HIV-1 Rev protein and its cellular export receptor CRM1 (exportin 1), as already reviewed in some detail (POLLARD and MALIM 1998; CULLEN 1998; STUTZ and ROSBASH 1998). In HIV-1-infected cells, Rev contributes to the switch from early to late gene expression. During the early phase of infection several viral regulatory proteins, including Rev, are synthesized from fully spliced viral mRNAs. These are generated by alternative splicing from a full-length primary transcript derived from a single promoter within the 5' long terminal repeat. Splicing, however, is inefficient, and unspliced as well as partially spliced viral messages accumulate in the nucleus because of host cell retention mechanisms that normally block the export of immature pre-RNAs still harboring functional introns. As the infection proceeds into the late phase, nuclear sequestration of the viral pre-mRNAs is counteracted by Rev, which efficiently promotes their export, subsequently allowing expression of the viral structural proteins. The nuclear export of partially spliced and unspliced viral RNAs requires the binding of Rev to the Rev responsive element (RRE) located within one intron of the HIV-1 genomic RNA, and the binding of CRM1 (exportin 1), a member of the importin/karyopherin-β superfamily of transport receptors, to the Rev NES, a nuclear export signal (NES) consisting of four critically spaced leucine residues on the Rev protein (reviewed in POLLARD and MALIM 1998; GÖRLICH and KUTAY 1999). It is assumed that RNA/Rev/CRM1-complexes, including Ran-GTP, migrate to the nuclear face of the NPC, which may involve interactions of CRM1 with FG-repeat-containing nucleoporins. Finally, the complex is translocated through the NPC to the cytoplasm.

As for most small DNA viruses and complex retroviruses, almost all adenovirus primary transcripts are differentially spliced and polyadenylated to achieve the required pattern of viral protein synthesis from the limited number of promoters. In addition, the pattern of differential splicing changes during the course of infection, in general, shorter RNAs being produced from early and late genes at later points of infection as a result of splicing out larger introns (reviewed in IMPERIALE et al. 1995). It is therefore not surprising that regulated processing of adenovirus primary transcripts is a tightly controlled process that involves temporally coordinated changes in the use of different splice sites and poly (A) sites to produce the full spectrum of translatable transcripts needed at defined stages of the viral life cycle. Since viral mRNA processing relies almost entirely upon the host cell machinery, regulation relies on viral functions occurring at all levels of the processing pathway. An important consequence of the alternative splicing process is that many viral transcripts retain unused splice sites and, in some cases, contain potential intron sequences in their 5'- or 3'-noncoding regions. As mentioned before, intron-containing pre-mRNAs are normally retained in the nucleus by the interaction of essential splicing factors (often referred to as commitment factors) until they are either spliced to completion or degraded (LEGRAIN and ROSBASH 1989). This would present adenovirus, and also complex retroviruses, with the problem that most viral mRNAs retaining characteristics of immature transcripts reach the cytoplasm inefficiently, because they are recognized, retained and subsequently degraded by host nuclear retention mechanisms. Adenovirus, like retroviruses, has therefore evolved different but closely integrated posttranscriptional strategies by which its RNAs circumvent nuclear sequestration and succeed in being exported to the cytoplasm. As will be discussed in more detail in the following sections, nuclear mRNA retention is effectively resolved by virus-encoded components that: (1) initiate a program of nuclear reorganization that promotes selective mRNA export, (2) operate as splicing regulators with positive effects on viral mRNA stability and/or RNA commitment to transport pathways, (3) facilitate viral mRNP release from the nuclear matrix, and (4) function as export adaptors similar to HIV-1 Rev.

3.2 Evidence for Regulated RNA Export in Adenovirus-Infected Cells

Evidence for regulation of viral mRNA export in adenovirus-infected cells comes from genetic studies of virus mutants. Much of this information has been reviewed in detail previously (IMPERIALE et al. 1995; LEPPARD 1997, 1998) and will be summarized only briefly here. Early studies on the Ad2/Ad5 replicative cycle in HeLa cells showed that at late times of infection E1B-55kDa facilitates the cytoplasmic accumulation of viral late transcripts while simultaneously inhibiting the export of most cellular mRNAs (BELTZ and FLINT 1979; BABISS et al. 1985; PILDER et al. 1986; WILLIAMS et al. 1986; LEPPARD and SHENK 1989). In addition, E1B-55kDa also blocks host protein synthesis, probably by mechanisms unrelated to the inhibition of cellular mRNA export (BABICH et al. 1983). Consequently, viral mutants that fail to express the E1B-55kDa protein show severe defects in late gene expression, viral replication and shut-off of host cell protein synthesis in the cytoplasm. A similar, but not identical, phenotype was observed for viruses bearing mutations in the gene encoding the 34-kDa protein from E4orf6 (HALBERT et al. 1985; WEINBERG and KETNER 1986). Mutants that lack E4orf6 show a reduction in late viral mRNA in the nucleus as well as a defect in the cytoplasmic accumulation of this RNA (WEINBERG and KETNER 1986; SANDLER and KETNER 1989; BRIDGE and KETNER 1990). Since both viral proteins associate in infected cells (SARNOW et al. 1984), it was concluded that a complex of E1B-55kDa and E4orf6 is the active component that regulates viral gene expression posttranscriptionally at the level of RNA transport (BABISS and GINSBERG 1984; BABISS et al. 1985; HALBERT et al. 1985; PILDER et al. 1986; WILLIAMS et al. 1986). This idea was further supported by the characterization of double mutants that showed a reduction in late protein synthesis essentially identical to that observed for viruses bearing either single mutation (CUTT et al. 1987; BRIDGE and KETNER 1990). In the original genetic analysis of region E4, no phenotype was associated with mutants unable to express the 11-kDa protein from open reading frame 3 (E4orf3), since, unlike E4orf6 discussed above, disruption of E4orf3 did not reduce virus viability in HeLa cells (HALBERT et al. 1985). However, double mutants of E4orf3 and E4orf6 displayed severe defects in late gene expression and replication suggesting that these proteins are mutually redundant in viral lytic infection by having compensatory functions (BRIDGE and KETNER 1989; HUANG and HEARING 1989). This phenotype was consistent with the reduction of nuclear and cytoplasmic late viral mRNAs and proteins as observed with E1B-55kDa and/or E1B-55kDa/E4orf6 mutants. Studies of other mutants deficient in both E1B-55kDa and E4orf3 gave similar results (BRIDGE and KETNER 1990). Collectively, these data suggested that three early adenovirus proteins are required for maximal cytoplasmic accumulation of late viral mRNAs and implicated some redundancy of function between E1B-55kDa/E4orf6 and E4orf3 in the control of viral late mRNA processing and late mRNA transport.

3.3 Regulated Viral RNA Export Is Linked to Alterations in Nuclear Organization

E1B-55kDa, E4orf3 and E4orf6 do not show any extensive primary sequence homology with other proteins known to be involved in posttranscriptional regulation, although both E1B-55kDa and E4orf6 contain sequence motifs which may link these proteins to RNA binding and nuclear export (see Sects. 3.5, 3.6). Consistent with their roles in late viral mRNA metabolism, both E4orf3 and E4orf6 are nuclear proteins. Whereas E4orf6 exhibits an even distribution throughout the nucleoplasm (CUTT et al. 1987), E4orf3 has been found to associate with and cause reorganization of nuclear domains known as Promyelocytic leukemia (PML) oncogenic domains (PODs; CARVALHO et al. 1995; DOUCAS et al. 1996). PODs, alternatively known as nuclear dots (NDs), nuclear domains 10 (ND10), PML-associated bodies, Kr-bodies, or nuclear bodies (NBs), are nuclear sub-structures closely attached to the nuclear matrix that have been associated with various aspects of cell growth regulation and the life cycles of several viruses (reviewed in STERNSDORF et al. 1997; MAUL 1998; SEELER and DEJEAN 1999). As opposed to the E4 proteins, E1B-55kDa exhibits a complex distribution in the cytoplasm and nucleus of virus-infected cells (SARNOW et al. 1982a; SMILEY et al. 1990; ORNELLES and SHENK 1991). As the following discussion will demonstrate, this diverse localization depends on progressive alterations of existing nuclear structures and on the formation of new virus-induced structures that are probably a prerequisite for selective viral mRNA export.

The initial phase of adenovirus infection is characterized by the appearance of small, electron-dense nuclear inclusions (replicative foci) which are located in close proximity to POD antigens (ISHOV and MAUL 1996). The replicative foci contain single- and double-stranded DNA as well as spliced and non-spliced viral RNAs and represent the sites of early viral transcription and replication (reviewed in BRIDGE and PETTERSSON 1995). As the early phase proceeds, E4orf3 induces the reorganization of PODs into track-like formations (also called spicules) and itself becomes concentrated in the nuclear tracks (Fig. 2; CARVALHO et al. 1995; DOUCAS et al. 1996). Although E4orf3 is sufficient for induction of track formation in transfected and virus-infected cells, it appears that at higher multiplicities of infection viral proteins other than E4orf3 may also have a limited ability to promote POD redistribution (DOUCAS et al. 1996). Possible candidates include E1B-55kDa, which associates with intact PODs (DOUCAS et al. 1996), and E1A, which localizes with reorganized POD components in virus-infected cells (CARVALHO et al. 1995). Immediately upon POD reorganization, E1B-55kDa becomes concentrated in the

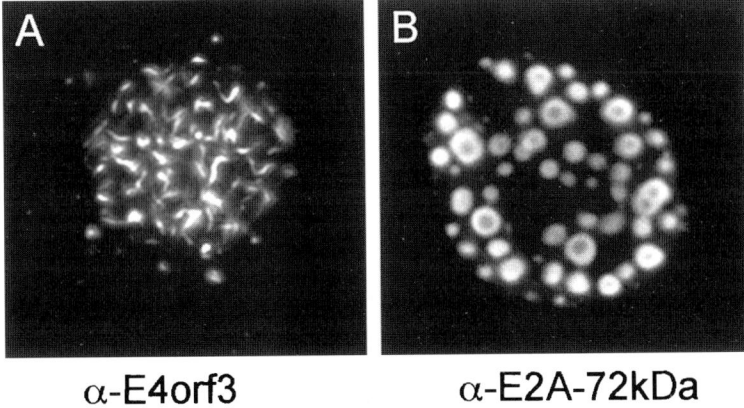

α-E4orf3   α-E2A-72kDa

Fig. 2A,B. Nuclear tracks and viral transcription/replication centers in Ad5 wild-type infected cells. **A** Ad5 E4orf3 colocalizes with nuclear tracks. A549 cells were infected with *wt*300 virus and stained with anti-E4orf3 monoclonal antibody 6A11 (NEVELS et al. 1999). **B** Late viral transcription/replication centers. A549 cells were infected with *wt*300 virus and stained with anti-E2A-72kDa (DBP) antibody B6–8. (The images were kindly provided by Birgitt Täuber)

nuclear tracks (DOUCAS et al. 1996; KÖNIG et al. 1999). This localization is strictly dependent on E4orf3 (KÖNIG et al. 1999) and requires the interaction of E1B-55kDa with the E4orf3 protein (LEPPARD and EVERETT 1999).

Following the onset of viral DNA synthesis, the replicative foci become enlarged and are separated into two distinct areas: a single-stranded DNA accumulation site and a peripheral zone where viral chromosomes are replicated and transcribed (PUVION DUTILLEUL and PICHARD 1992; BRIDGE and PETTERSSON 1996). Also, concomitant with the activation of late viral gene expression, splicing factors are redistributed from their normal localization in interchromatin granules (IGs) and coiled bodies to the peripheral zone of replication centers (JIMÉNEZ-GARCIA and SPECTOR 1993; BRIDGE et al. 1993; POMBO et al. 1994; REBELO et al. 1996). As the infection proceeds into the late phase and the full spectrum of late viral mRNAs are produced, large viral inclusion bodies known as viral genome storage sites appear, which are surrounded by IGs and replicative foci comprising both single-stranded DNA accumulation sites and the peripheral replicative zone (Fig. 2; BRIDGE and PETTERSSON 1996). At this stage splicing factors are shifted back to the IGs, which become enlarged (IG clusters) and now contain late viral RNAs enriched in exon sequences (BRIDGE et al. 1996; ASPEGREN et al. 1998). Interestingly, in the same time-frame, the association of E1B-55kDa with nuclear tracks is lost. E1B-55kDa becomes more diffusely distributed throughout the nucleoplasm (GOODRUM et al. 1996; KÖNIG et al. 1999) and some of the E1B-55kDa molecules along with reorganized POD components move to the periphery of viral inclusions bodies (ORNELLES and SHENK 1991; DOUCAS et al. 1996). It appears that the release of E1B-55kDa from the track structures requires the E4orf6 protein, which associates with E1B-55kDa in virus-infected cells (SARNOW et al. 1984; CUTT et al. 1987; RUBENWOLF et al. 1997) and presumably competes with E4orf3 for

E1B-55kDa binding (KÖNIG et al. 1999; LEPPARD and EVERETT 1999). At present it is not clear whether E4orf6 is also required to direct E1B-55kDa to the viral nuclear inclusion bodies (ORNELLES and SHENK 1991; KÖNIG et al. 1999) and whether E4orf6 colocalizes with E1B-55kDa in the virus-induced structures. Nonetheless, the relocation of E1B-55kDa to the active sites of late viral transcription and replication appears to be a key step in the selective export of viral late mRNAs and may also be required to block the export of most host cell transcripts (see Sect. 3.5). Apparently, this key localization is achieved by a coordinated series of sequential events. These may involve the initiation of viral DNA replication and activation of late gene expression at the MLTU as well as a variety of protein interactions, including the timely coordinated formation of E1B-55kDa/E4orf3 and/or E1B-55kDa/E4orf6 complexes. The observation that POD-associated components are recruited to the viral nuclear inclusions suggests that E4orf3-induced POD reorganization may also play a role in this process. It should be mentioned, however, that the functional significance of POD disruption with respect to lytic viral replication is unclear, mainly because E4orf3 mutants display no obvious phenotype in HeLa cells (HALBERT et al. 1985) and other standard tissue culture cell lines. The finding that inhibition of POD reorganization severely delays adenovirus replication (DOUCAS et al. 1996) suggests that E4orf3 accelerates the accretion of otherwise limiting cellular factors for late viral gene expression and viral DNA replication. Since PODs are intimately linked to the formation of viral replication compartments and efficient viral replication by several viruses (reviewed in STERNSDORF et al. 1997; MAUL 1998), it seems likely that modulation of the POD structure and POD-associated proteins plays a more significant role in infection processes in nature or in primary cell strains, and presumably at lower multiplicities of infection.

3.4 E4orf3 and E4orf6 Maintain the Stability of Major Late RNAs in the Nucleus

Direct evidence for a role of E4orf3 and E4orf6 in posttranscriptional regulation of viral RNA biogenesis originates from studies implicating both proteins in the stable nuclear accumulation of late mRNAs derived from the MLP (HALBERT et al. 1985; WEINBERG and KETNER 1986; SANDLER and KETNER 1989). Mutants unable to express either one or both E4 proteins were found to exhibit an enhanced rate of viral late transcript degradation in the nucleus, indicating that E4orf3 and E4orf6 act in parallel to improve the stability of late viral RNAs (BRIDGE and KETNER 1989; SANDLER and KETNER 1989). Subsequent analyses showed that E4orf3 and E4orf6 posttranscriptionally regulate late TPL splicing (NORDQVIST and AKUSJÄRVI 1990; ÖHMAN et al. 1993; NORDQVIST et al. 1994). It was shown that E4orf3 functions as an exon inclusion factor while E4orf6 promotes exclusion of the *i* leader exon in assays of MLTU splicing (Fig. 1) either in vitro or in vivo (ÖHMAN et al. 1993; NORDQVIST et al. 1994). These observations led to the proposal that both E4 proteins operate in a similar fashion to cellular factors involved in regulation of alternative splicing, such as ASF/SF2 (E4orf3) and hnRNP A1 (E4orf6) (NORDQVIST et al. 1994), and that the redundant functions of E4orf3 and E4orf6 in nuclear RNA stabilization may be directly linked to these activities (LEPPARD 1997, 1998).

The proposition that E4orf3 and E4orf6 function as virus-encoded splicing regulators further suggests that both early proteins may modulate the commitment of viral mRNAs derived from the MLP in cellular splicing pathways. The commitment machinery functions through identifying and defining introns during the initial steps in the normal splicing pathway, leading to the assembly of further small nuclear ribonucleoprotein particles (snRNPs) and splicing factors on the RNA (LEGRAIN and ROSBASH 1989). In addition, splicing commitment factors, such as the U1 snRNP and members of the serine-arginine-rich (SR) class of splicing factors, retain incorrectly or partially spliced transcripts in the nucleus until they are either spliced to completion or degraded (LEGRAIN and ROSBASH 1989; CHANG and SHARP 1989; HAMM and MATTAJ 1990). Conceivably, the modulation of early splice site selection (commitment step) within the TPL and/or enhancement of splicing in general could potentially prevent nuclear degradation of major late mRNAs and thus maintain the pool of late viral transcripts available for mRNA transport.

Finally, it is important to note that the significance of the E4orf3/E4orf6 splicing function in late viral gene expression remains to be fully defined, since it is not clear how two proteins with opposite activities in TPL assembly can complement for one another in major late RNA stabilization. This raises the possibility that other or additional functions contribute to the effects of these early proteins in viral mRNA stabilization and late gene expression. These include the E4orf3-induced relocation of POD-associated components and the ability of E4orf6 to inhibit DNA double-strand break repair through interaction with the catalytic subunit of the cellular DNA-dependent protein kinase (BOYER et al. 1999; NICOLÁS et al. 2000). Although the latter activity might ultimately be linked to preventing aberrant concatemer formation of viral chromosomes during viral DNA replication (WEIDEN and GINSBERG 1994), it is certainly possible that the absence of these functions may indirectly affect nuclear stability of viral mRNAs and thus late viral gene expression. In addition, a third E4 gene product encoded in open reading frame 4 (E4orf4) has recently been shown to regulate the temporal shift in L1 52/55-IIIa alternative RNA splicing (Fig. 1) in vitro and in vivo (KANOPKA et al. 1998). This early protein affects several cellular processes including down-regulation of virally induced signal transduction, regulation of gene expression and induction of p53-independent apoptosis (MÜLLER et al. 1992; KLEINBERGER and SHENK 1993; LAVOIE et al. 1998; SHTRICHMAN and KLEINBERGER 1998; SHTRICHMAN et al. 1999), processes which may also modulate the effects of E4orf3/E4orf6 in viral DNA replication (BRIDGE et al. 1993). Most of these activities correlate with the binding of E4orf4 to protein phosphatase (PP)2A (KLEINBERGER and SHENK 1993). At the posttranscriptional level, E4orf4 appears to regulate alternative splicing by reducing the RNA-binding capacity of HeLa SR proteins due to their dephosphorylation by PP2A. The importance of this process for efficient late viral gene expression is unclear, because E4orf4, like E4orf3, is not essential for lytic growth in HeLa cells (HALBERT et al. 1985). Clearly further work is needed to fully understand a possible relationship between the effects of these E4 proteins and viral DNA synthesis, regulated viral mRNA splicing and possibly viral mRNA export.

3.5 E1B-55kDa/E4orf6 Counteract Nuclear Retention of Viral RNAs

The initial studies with E1B-55kDa mutants demonstrated that this adenovirus protein is required for the efficient cytoplasmic accumulation of transcripts derived from the MLP in late-infected cells but not for the export of viral mRNAs transcribed from early viral promoters at early times (BABISS and GINSBERG 1984; PILDER et al. 1986; WILLIAMS et al. 1986). These observations were refined by LEPPARD (1993), who showed that the export of mRNAs transcribed from other viral promoters (IX, IVa2, and E2-L) activated during the late phase of infection is also dependent on this E1B function. These results, along with the findings that cellular mRNAs transcribed from recombinant viral chromosomes are moved to the cytoplasm late after infection (GAYNOR et al. 1984; HEARING and SHENK 1985), indicated that the selective transport observed in late adenovirus-infected cells is generally not based on recognition of a common RNA sequence. Instead, it appears that transcriptional activation of viral genes during the late phase of infection determines the E1B-55kDa-dependent export of the relevant mRNAs. This conclusion is supported by the observation that cytoplasmic accumulation of L1–52/55 mRNA, synthesized from the major late transcription unit, is strongly E1B-dependent during the late, but not the early phase of infection (LEPPARD 1993) and the recent finding that RNA products from early region 1A (E1A) that are not specifically activated during the late phase are not selectively exported (YANG et al. 1996). In addition, some cellular transcripts escape the virus-induced inhibition of host cell mRNA transport. These include the heat shock protein 70, β-tubulin, and the interferon-inducible Mx-A and 6–16 mRNAs (MOORE et al. 1987; YANG et al. 1996). Significantly, efficient transport of this subset of cellular messages correlates with transcriptional activation of the corresponding genes in the late phase and, as with viral late mRNAs, requires the E1B-55kDa protein (YANG et al. 1996). Taken together these data strongly indicate that transcriptional activation in late-infected cells determines the export of cellular and viral transcripts and suggest that these mRNAs (viral or cellular) are transported through the same E1B-55kDa/E4orf6-dependent export pathway.

The site of action of the E1B-55kDa protein in the transport pathway was defined more closely by monitoring the movement of newly synthesized RNAs through a series of biochemically defined nuclear subfractions. In the absence of E1B-55kDa, late viral RNAs displayed a reduced ability to chase from a nuclear matrix fraction and failed to accumulate in a nuclear downstream compartment (the soluble, nuclear fraction) because of enhanced degradation within the nucleus

(LEPPARD and SHENK 1989). In a similar study employing wild-type Ad2, it was shown that cellular mRNAs failed to accumulate in the soluble, nuclear fraction, whereas viral mRNAs were efficiently recovered within this operationally defined compartment (DENOME et al. 1989). From these observations it was concluded that E1B-55kDa facilitates an early rate-limiting step in the movement of viral mRNAs on the nuclear matrix, resulting in improved transport to the cytoplasm, while simultaneously restricting the same process for cellular transcripts. Detailed analysis of the levels of several MLP-derived mRNA species revealed that the cytoplasmic accumulation of the various differentially spliced products of the MLTU are not dependent on E1B-55kDa to the same extent (LEPPARD 1993). Rather, accumulation of the longest mRNA species of the 3′-coterminal set strongly depended on E1B-55kDa, whereas the shortest mRNA in each family, which becomes disproportionately abundant, exhibited less or no dependence on this activity (LEPPARD 1993). This observation suggests that the strong dependence of late viral mRNA accumulation on E1B-55kDa function correlates with the presence of unused splice sites and potential intron sequences within these late transcripts. It was therefore proposed that E1B-55kDa counteracts a host cell nuclear retention system on the nuclear matrix that prevents the efflux of immature transcripts from the nucleus (LEPPARD 1993). This idea gained subsequent support by the finding that the cytoplasmic accumulation of one partially spliced mRNA species from E4 was uniquely dependent on expression of both E1B-55kDa and E4orf6 (DIX and LEPPARD 1993). This mRNA belongs to the late class of E4 transcripts and retains an intron from the 5′ half of the E4 pre-mRNA (DIX and LEPPARD 1993).

Although there is no evidence for any sequence motif common to all E1B-55kDa/E4orf6-dependent viral and cellular mRNAs, recent data have shown that the presence of the TPL can increase the efficiency of processed RNA export (HUANG and FLINT 1998). Significantly, higher levels of reporter RNA transcripts whose expression was activated late from a recombinant adenovirus accumulated in the cytoplasm if the TPL was also present. This observation might explain why the cytoplasmic accumulation of mRNAs transcribed from IX, IVa2 and E2-L promoters (which do not contain the TPL) are less dependent on E1B-55kDa than MLTU-derived transcripts (LEPPARD 1993). However, it is not yet clear whether this TPL-mediated effect on mRNA export requires the E1B-55kDa protein. In a similar investigation, the TPL simultaneously decreased the nuclear half-life and increased the cytoplasmic half-life of a thymidine kinase-specific reporter mRNA (MOORE and SHENK 1988). Although the impact on RNA export was not examined in this study, these effects were found to be independent of E1B-55kDa function.

A number of speculative (non-exclusive) proposals can be made at this point concerning the mechanism by which E1B-55kDa/E4orf6 promotes viral mRNA release from the retaining RNPs. The implication that viral mRNA export is dependent on residual splicing sites in different viral mRNAs suggests that proteins which regulate nuclear retention and release of pre-mRNAs may be targets for the E1B-55kDa/E4orf6 complex. In addition, the viral proteins might directly interact with RNA via some feature of the unused splice sites or intron sequences. One attempt to address this point examined the binding of E1B-55kDa to several viral RNA species known to be strongly dependent on E1B-55kDa export function (HORRIDGE and LEPPARD 1998). In vitro studies with bacterially expressed wild-type and mutant proteins revealed that E1B-55kDa binds non-specifically to these RNAs. The protein region implicated in RNA-binding displays limited homology to other RNA-binding proteins and maps to the central region of E1B-55kDa (Fig. 3; HORRIDGE and LEPPARD 1998), shown previously to be necessary for efficient viral replication (YEW et al. 1990). Although it is hard to envisage that such an activity contributes to the specificity of viral RNA export it is possible that E1B-55kDa RNA binding is regulated or modulated in virus-infected cells, for example, by other cellular and viral proteins such as E4orf3 and E4orf6 and/or posttranslational modifications.

Another hypothesis is that E1B-55kDa/E4orf6 operates through mechanisms that remove or inactivate cellular retention factors and/or target cellular proteins actively involved in the release of mRNAs from the retaining RNPs. Such a mechanism is compatible with the idea that the E1B-55kDa/E4orf6 complex interacts with a limiting cellular factor required for cytoplasmic accumulation of mRNAs and directs it to the periphery of the transcriptionally active viral inclusion bodies (ORNELLES and SHENK 1991). Considered in the context of this model, the sequestration of a positively acting nuclear RNA export factor into the viral transcription/replication centers could explain the ability of the viral protein complex to inhibit the release of RNAs transcribed from cellular locations (DENOME et al. 1989) while simultaneously stimulating accumulation of transcripts derived from the viral chromosome. Since the accumulation of spliced late transcripts in IG clusters correlates with efficient export from the nucleus (BRIDGE et al. 1996; ASPEGREN et al. 1998), such a factor could act on viral mRNAs arriving or already present in these defined compartments. Recently, a cellular protein called E1B-AP5 was identified that binds to Ad5 E1B-55kDa in vitro and in virus-infected cells (GABLER et al. 1998). E1B-AP5 is a nuclear RNA-binding protein that is

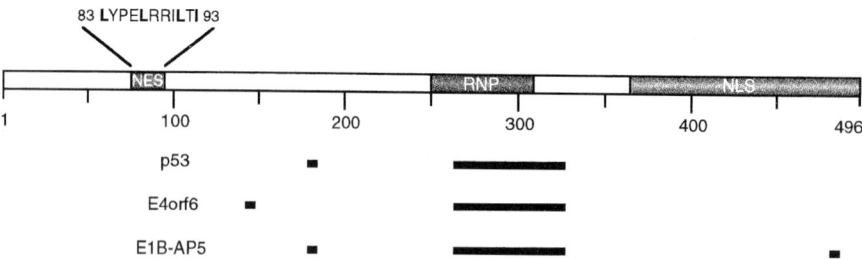

Fig. 3. Domains and motifs in the Ad5 E1B-55kDa protein. The amino acid sequence of the leucine-rich NES (residues 83–93) is indicated *above*. *RNP* denotes the region implicated in non-specific RNA-binding (HORRIDGE and LEPPARD 1998). The location of the nuclear localization signal (*NLS*) of Ad5 E1B-55kDa has been mapped to the carboxy-terminal region (residues 364–496) (KRÄTZER et al. 2000). The p53, E4orf6 and E1B-AP5 interaction domains (*black bars, below*) are shown relative to their positions on the E1B-55kDa polypeptide and were defined by YEW et al. (1992), RUBENWOLF et al. (1997), and GABLER et al. (1998)

related to hnRNPs from several species. The most significant similarity is with the central part of hnRNP-U/SAF-A nuclear matrix proteins from human, rat and chicken, which bind to RNA as well as to scaffold attachment regions (KILEDJIAN and DREYFUSS 1992; FACKELMAYER et al. 1994; VON KRIES et al. 1994). At present, three lines of evidence indicate that E1B-AP5 provides functions in viral and cellular mRNA processing and/or transport. First, the same regions on the E1B-55kDa protein that mediate binding to E1B-AP5 are crucial for normal virus growth (Fig. 3; YEW et al. 1990; RUBENWOLF et al. 1997). Second, high levels of the E1B-AP5 protein stimulate the export of viral late transcripts and simultaneously prevent the block of host cell mRNA export (GABLER et al. 1998). Third, E1B-AP5 can bind to the amino-terminal domain of TAP, an essential RNA export mediator that may bridge the interaction between specific RNP export substrates and the NPC (BACHI et al. 2000). We also note that the central region of E1B-AP5 shows substantial homology to the GTP-binding domain of Ran (GABLER et al. 1998). Given the assumption that E1B-AP5 possesses a GTPase activity, it might accelerate the release of viral mRNPs from the retaining matrix compartments, such as IG clusters. These data are compatible with the hypothesis that E1B-55kDa facilitates the cytoplasmic accumulation of viral transcripts by binding to a host factor that may promote RNP release (ORNELLES and SHENK 1991; LEPPARD 1993). However, the molecular mechanism by which E1B-AP5, along with E1B-55kDa, affects viral and cellular mRNA transport remains to be elucidated.

3.6 E1B-55kDa and E4orf6 Are Viral Shuttle Proteins

The finding that the Ad5 E1B-55kDa/E4orf6 complex continuously shuttles between the nuclei of heterokaryons formed between cotransfected HeLa and non-transfected mouse Balb-c 3T3 cells provides further evidence for a more direct role of E1B-55kDa and E4orf6 in viral mRNA export. The nuclear export function was found to be mediated by a leucine-rich NLS of the Rev-type located on the E4orf6 protein which functions autonomously when fused to a glutathione-S-transferase (GST) carrier protein (Figs. 4, 5; DOBBELSTEIN et al. 1997). Furthermore, export of E4orf6 can be blocked by an HTLV-1 Rex-derived competitive inhibitor and conversely wild-type E4orf6 inhibits Rev-mediated transport. Interestingly, when expressed alone E4orf6 does not move from the human to the murine nucleus. This nuclear retention can be reversed by cotransfection of E1B-55kDa or by mutations within the carboxy-terminal region of E4orf6 that is rich in arginine residues and overlaps one of the putative E4orf6 NLSs (GOODRUM et al. 1996; DOBBELSTEIN et al. 1997; ORLANDO and ORNELLES 1999; NEVELS et al. 2000). A peptide of this region (Fig. 4) displays the characteristics of an amphipathic α-helix (ORLANDO and ORNELLES 1999), and mutants of E4orf6 with disruptions of this α-helix fail to target E1B-55kDa to the nucleus (ORLANDO and ORNELLES 1999). Based on these observations, it was proposed that E4orf6 contains a nuclear retention-like sequence (NRS) in the carboxy-terminal region that is dominant over the NLSs located at the amino- and carboxy-termini. A direct or functional

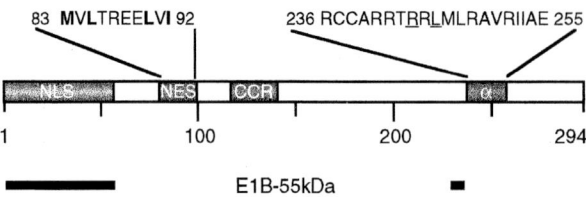

Fig. 4. Domains and motifs in the Ad5 E4orf6 protein. The amino acid sequences of the leucine-rich NES (residues 83–92) and the arginine-rich amphipathic α helix (α: residues 239–253) are indicated *above*. The residues of the RXL motif (GRIFMAN et al. 1999) within the α helix are *underlined*. *CCR* denotes the highly conserved cysteine-rich motif (NEVELS et al. 2000). The location of the amino-terminal NLS is indicated (ORLANDO and ORNELLES 1999; NEVELS et al. 2000); the second NLS overlaps with the amphipathic α-helix (ORLANDO and ORNELLES 1999; NEVELS et al. 2000; WEIGEL et al. 2000). The E1B-55kDa interaction domains (*black bars*, *below*) are shown relative to their positions on the E4orf6 polypeptide, and were defined by RUBENWOLF et al. (1997) (residues 1–58) and WEIGEL et al. (2000) (residues 225–232)

```
HIV-1 Rev        L P P . L E R . L T L
HTLV-I Rex       L S A Q L Y S S L S L
PKI              L A L K L A G . L D I
Gle1p            L P . . L G K . L T L
Ad5 E4orf6       M V . L T R E E L V I
Ad5 E1B-55kDa    L Y P E L R R I L T I
consensus        L X_{2-3}  λ  X_{2-3}  L X L/I
```

Fig. 5. Sequence comparison of leucine-rich nuclear export signals. Alignment of NESs from HIV-1 Rev, HTLV-I Rex, inhibitor of cAMP-dependent protein kinase (*PKI*), Gle1p, Ad5 E4orf6 and Ad5 E1B-55kDa. The *dots* in the sequence indicate gaps, X represents any amino acid and λ indicates amino acids with bulky hydrophobic side chains. The NESs of Rev, Rex, PKI, Gle1p and the consensus sequence were derived from POLLARD and MALIM (1998)

interaction of the amphipathic α-helix with E1B-55kDa appears to counteract the nuclear restriction imposed on E4orf6, whereupon both proteins can move from the nucleus to the cytoplasm possibly through a Rev NES-dependent export pathway (DOBBELSTEIN et al. 1997). However, it should be mentioned that it is not yet clear whether E4orf6 requires E1B-55kDa for shuttling. Ornelles and coworkers have reported that E4orf6 alone shuttles between the nuclei in heterokaryons of rat and HeLa cells as well as within heterokaryons of HeLa and rat cells (GOODRUM et al. 1996; ORLANDO and ORNELLES 1999). The latter observation is consistent with data from FISCHER et al. (1999), who also observed shuttling of E4orf6 between nuclei of HeLa cells and murine Balb-c 3T3 cells in the absence of E1B-55kDa.

The contribution of the E4orf6 NES and the α-helix to lytic virus growth was recently analyzed by combined transfection/infection experiments in HeLa cells (ORLANDO and ORNELLES 1999; WEIGEL and DOBBELSTEIN 2000). Mutations in the NES considerably reduced virus yield, viral DNA synthesis, late hexon protein production and cytoplasmic accumulation of L5 mRNA. Significantly, virus yield and hexon synthesis was also diminished when an export competitor derived from the HTLV-1 Rex protein was cotransfected with wild-type E4orf6, while a mutant construct of the Rex NES had no effect in these assays (WEIGEL and DOBBELSTEIN 2000). Mutations within or preceding the α-helical region substantially reduced viral replication, hexon synthesis, and L5 mRNA export and abolished the ability of E4orf6 to direct E1B-55kDa to the nucleus as well as binding to E1B-55kDa from 293 cells (ORLANDO and ORNELLES 1999; WEIGEL and DOBBELSTEIN 2000). Together, these data indicate that E4orf6 contains at least two functional domains which support lytic virus growth: an arginine-faced amphipathic α-helix critical for nuclear retention and interaction with E1B-55kDa, presumably through a primate cell-specific factor (GOODRUM et al. 1996), and a leucine-rich NES involved in cytoplasmic accumulation of viral late mRNAs. In addition, it was recently suggested that the amphipathic α-helix of E4orf6 contains a putative RXL motif (Fig. 4) essential for augmentation of adeno-associated virus vector (rAAV) transduction (GRIFMAN et al. 1999). As will be discussed (Sect. 5), this motif may link E4orf6 to an unanticipated role in the control of cell cycle progression and protein degradation.

Curiously, recent work has demonstrated that the Ad5 E1B-55kDa protein itself shuttles efficiently in the absence of E4orf6 and that trafficking is also independent of Mdm2 and p53 (KRÄTZER et al. 2000). Similar to E4orf6, E1B-55kDa export is mediated by a previously postulated leucine-rich NES of the HIV-1 Rev-type (Figs. 3, 5; LIAO et al. 1999) and can be blocked by point mutations within the E1B NES or by export competitors such as Rev and Rex, as well as leptomycin B (LMB). As opposed to the E4orf6 protein, E1B-55kDa does not inhibit the activity or the trafficking of the retroviral shuttle proteins Rev and Rex. This indicates that the retroviral proteins display higher affinities for common export components that may be shared between E1B-55kDa and Rev/Rex. The NES of E1B-55kDa, located at the amino-terminus (Fig. 3), mediates rapid nuclear export when fused to heterologous proteins (GST or GST-green fluorescent protein) and microinjected into HeLa cell nuclei. However, contrary to the previous work on E4orf6 shuttling, no substantial export was detected in these assays using the E4orf6-NES fused to GST or green fluorescent protein (GFP) or a mutated E4orf6 NES that more closely matched the postulated consensus sequence for leucine-rich NESs (KRÄTZER et al. 2000). The reason for these apparent discrepancies remains to be determined. Nevertheless, the finding that E1B-55kDa itself shuttles in the absence of E4orf6 suggests that many of the considerations described above deserve re-evaluation.

Despite these inconsistent observations, it seems likely that E1B-55kDa/E4orf6 may function as viral export adaptors that links the viral mRNA cargo to nuclear export receptors. Because E4orf6 and E1B-55kDa trafficking is mediated by leucine-rich NESs and can be blocked by the export competitors Rev and Rex, it is conceivable that late viral mRNA export may occur through the interaction with CRM1, which appears to be an essential export receptor for most Rev NES-containing proteins. We note that Rev expressed from a recombinant adenovirus genome modestly enhances export of RRE-containing major late RNAs in the absence of E1B-55kDa (WILLIAMS and LEPPARD 1996). Although CRM1 has been implicated in nuclear RNA export of U snRNAs and 5S rRNA, it may not play a major role in cellular mRNA export (reviewed in MATTAJ and ENGLMEIER 1998). Instead, substantial evidence indicates that other cellular proteins, termed TAP (yeast Mex67p) and Gle1 (yeast Gle1p), which are distinct from the nuclear transport receptors of the β-related, Ran-binding protein family, are all critical components in mediating global nuclear mRNA export (reviewed in GÖRLICH and KUTAY 1999; NAKIELNY and DREYFUSS 1999). Thus, considered in the context of the reciprocal effects on viral and cellular mRNA export observed in late adenovirus-infected cells, E1B-55kDa/E4orf6 may use a CRM1-dependent pathway for nuclear export of late viral mRNAs while simultaneously modulating the activity and/or distribution of essential cellular components involved in general mRNA transport pathways, such as TAP and/or Gle1. This concept is intriguing given the observation that the E1B-associated protein E1B-AP5 can bind to TAP (BACHI et al. 2000) and that NLS-tagged Ad5 E1B-55kDa blocks bulk poly(A)$^+$ export in *Saccharomyces cerevisiae* (LIANG et al. 1995). Significantly, the inhibition of cytoplasmic poly(A)$^+$ RNA accumulation is dependent on the level of expression and the NLS (LIANG et al. 1995), which indicates that nuclear NLS-E1B-55kDa modulates the activity of a saturable factor essential for mRNA export in yeast. Since NLS-tagged Ad5 E4orf6 has no effect on poly(A)$^+$ RNA export in yeast, one might predict that the late functions required for inhibition of cellular mRNA export in virus-infected cells are encoded predominantly in the E1B-55kDa polypeptide.

The proposition that selective cytoplasmic accumulation of viral late transcripts is linked to the leucine-rich NES-dependent export pathway implies that viral mRNA transport is distinct from the general mRNA export pathways. A similar situation has been described for the export of heat shock (hs) mRNAs in *S. cerevisiae*. Hs RNA export in yeast is apparently independent of the Ran regenerating system and the hnRNP protein Npl3p, which is essential for general, but not for hs mRNA export (reviewed in COLE and HAMMELL 1998). More importantly, hs RNA export at 42°C absolutely requires the otherwise inessential FG-nucleoporin Rip1p (STUTZ et al. 1997; SAAVEDRA et al. 1997). Interestingly, overexpression of wild-type Rev partially blocks the export of hs mRNAs (SAAVEDRA et al. 1997), indicating that yeast Rip1p has some role in Rev NES-mediated nuclear export of hs RNPs. Consistent with this possibility, Rip1p, like the distantly related human Rip (hRip also called Rab or Rip/Rab), indirectly interacts with Rev NES through the export factor CRM1(p) (STUTZ et al. 1995; NEVILLE et al. 1997). Although the relationship between hs mRNA export and NES-dependent export is still unclear, these data imply that hs RNA export is probably distinct from general export pathways. Considering an analogous situation in adenovirus-infected cells, one might hypothesize that E1B-55kDa/E4orf6 usurps a CRM1/Rip-mediated export pathway similar to that/those induced and/or used under stress conditions for efficient and selective export of viral late RNPs. Such a model would also be compatible with the observation that the export of some cellular mRNAs transcribed following stress, such as heat shock (hsp70; MOORE et al. 1987; YANG et al. 1996) or virus infection (interferon-induced Mx-A and 6–16; YANG et al. 1996), are selectively exported by an E1B-55kDa-dependent mechanism (see Sect. 3.5). Clearly further work is needed to test the validity of such a model.

3.7 Requirement for E1B-55kDa and E4orf6 in Late Viral RNA Export Is Dependent on the Type of Host Cell

As described (Sect. 3.2), early analyses of the Ad2/Ad5 replicative cycle were performed primarily in HeLa cells, where viral replication is strictly dependent on E1B-55kDa and E4orf6 functions (for references, see Sect. 3.2). However, it was noticed that some E1B-55kDa and E4orf6 mutants grew productively in normal human embryonic kidney cells and some human tumor cell lines (HARRISON et al. 1977; BERNARDS et al. 1986; HALBERT et al. 1985; WILLIAMS et al. 1986; CUTT et al. 1987; SMILEY et al. 1990), indicating that E1B-55kDa and E4orf6 are required for viral replication in some host cells but not in others. Recently, researchers at ONYX Pharmaceuticals reported that an E1B mutant virus *dl*1520 (renamed ONYX-015 for commercial reasons), which lacks essentially all E1B-55kDa functions (BARKER and BERK 1987), replicates similarly to wild-type Ad5 in p53-negative human tumor cells but is attenuated in non-transformed primary cells (BISCHOFF et al. 1996; HEISE et al. 1997). Since the requirement for E1B-55kDa in neutralizing p53-induced growth arrest and apoptosis during infection is well known (reviewed in WHITE 1998; ROULSTON et al. 1999; and see Sect. 5), it was proposed that *dl*1520 preferentially replicates in tumor cells lacking functional p53. However, recent work by a number of investigators has clearly demonstrated that *dl*1520 and other Ad5 or Ad12 E1B-55kDa mutants replicate to near wild-type levels in many different p53-positive and p53-negative tumor cell lines or non-immortalized human primary cell strains. In others, lytic replication is severely restricted in a manner similar to HeLa cells (GOODRUM and ORNELLES 1997, 1998; ROTHMANN et al. 1998; TURNELL et al. 1999; HARADA and BERK 1999). Thus, p53 is not the host-range determinant of E1B-55kDa mutant virus growth, although it may impose a modest defect in viral DNA replication and late viral mRNA transport in cells expressing temperature-sensitive p53 (GOODRUM and ORNELLES 1998; HARADA and BERK 1999). Together, these observations demonstrate that substantial viral mRNA export and translation can occur in the absence of E1B-55kDa or E4orf6 in some host cells. This suggests that the replication and cytolytic properties of E1B-55kDa and E4orf6 mutants are determined by different genetic backgrounds in various cell lines and that non-restrictive cells compensate for E1B-55kDa and E4orf6 functions irrespective of the p53 status of the host cell.

At present it is unknown why E1B-55kDa and E4orf6 are required for viral replication in some host cells and not in others. Recent studies from TURNELL et al. (1999) indicate that the replicative capacity of E1B-55kDa mutants is enhanced in growing, actively cycling cells compared to quiescent cells. They suggested that cells actively progressing through the cell cycle provide cellular factors that permit mutant E1B-55kDa virus replication. Studies from Goodrum and Ornelles have shown that replication of both E1B-55kDa and E4orf6 mutant viruses is partially restored when restrictive HeLa cells are infected during S phase. Under these conditions transport of late mRNAs is increased and the infected cells synthesize more late viral proteins than cells infected with the mutant viruses during G1 phase (GOODRUM and ORNELLES 1997, 1998). By contrast, infection of cells during S phase with wild-type Ad5 results in a decrease of late mRNA export compared to G1-phase cells (GOODRUM and ORNELLES 1999). These observations indicate that viral mRNA transport is differentially regulated in response to the stage of the cell cycle at infection. Interestingly, as with synchronized cell populations, E1B-55kDa mutant replication is partially restored when restrictive cells are infected and maintained at 39°C. At the elevated temperature, late viral gene expression and therefore viral mRNA transport are only modestly reduced compared to that of wild-type virus (GOODRUM and ORNELLES 1998; HARADA and BERK 1999). Thus, S phase cells or infection at elevated temperatures may provide a common feature or factor that partially compensates for E1B-55kDa and E4orf6 late-phase functions. In this regard it is interesting to note that E4orf3 is required for enhanced replication of E4orf6 mutant viruses in cells infected during S phase (GOODRUM and ORNELLES 1999). Given the established role of E4orf3 in POD-disruption (see Sect. 3.3) and the observations that the POD structure is modulated during cell cycle progression and heat shock (reviewed in STERNSDORF et al. 1997, and references therein) one might speculate that POD-associated factors contribute to the virus infection during S phase or at elevated temperature. Such factors may participate directly in mRNA transport and/or translation (see Sect. 4) and could represent an intrinsic property of various non-restrictive cell lines that support mutant virus growth. We note that two POD components, PML and Sp100, localize in IGs (PUVION-DUTILLEUL et al. 1995), implying that POD-associated proteins may function in RNA metabolism. The observation that E1B-55kDa and E4orf6 mutant replication is restricted by the stage of the cell cycle further suggests that the late-phase functions of both factors are linked to components that participate in both nucleocytoplasmic transport of macromolecules and cell cycle regulation. Examples of proteins involved in both processes include RCC1 and the small GTP-binding protein Ran (reviewed in GÖRLICH and MATTAJ 1996). How these and other assorted proteins contribute to the regulation of late viral mRNA export remains to be determined.

4 Involvement of E1B-55kDa in Translational Control

As noted before (Sect. 2), the late phase of adenovirus infection is marked by the inhibition of cellular and preferential translation of major late mRNAs (BELTZ and FLINT 1979). This process is not dependent on the inhibition of host cell mRNA export but involves several virus-mediated translational control mechanisms that include the inactivation of initiation factor eIF-4F (see below), the viral late RNA-binding protein L4-100kDa, the TPL, and VAI RNAs, which prevent the activation and antiviral activities of the interferon-induced kinase DAI (reviewed in MATHEWS and SHENK 1991; ZHANG and SCHNEIDER 1993). In addition, virus mutants unable to express E1B-55kDa fail to shut off host protein synthesis or stimulate late viral mRNA translation (BABICH et al. 1983; BABISS and GINSBERG 1984; BABISS et al. 1985). This correlates with the inability of the mutant viruses to inactivate initiation factor eIF-4F, a cap-dependent RNA helicase required for translation of most capped mRNAs (HUANG and SCHNEIDER 1991). Factor eIF-4F is normally activated by phosphorylation of its cap-binding component eIF-4E. In late infection with wild-type virus, when cellular mRNA translation is blocked, eIF-4E becomes severely underphosphorylated, thus limiting its cellular activity (HUANG and SCHNEIDER 1991). Translation of viral late mRNAs, however, continues due to the presence of the TPL, which reduces or eliminates the requirement for eIF-4F (LOGAN and SHENK 1984; BERKNER and SHARP 1985; DOLPH et al. 1988, 1990). Interestingly, in HeLa cells infected with E1B-55kDa mutant virus *dl*338 (PILDER et al. 1986), eIF-4E remains phosphorylated (ZHANG et al. 1994) indicating that the adenovirus signal which induces eIF-4E dephosphorylation might be E1B-55kDa itself or another viral protein poorly expressed in the absence of the E1B-55kDa protein. The work from Schneider and coworkers (ZHANG et al. 1994) suggests that inhibition of eIF-4E is linked to an activity of one or more late viral polypeptides and that the failure of E1B-55kDa mutants to reduce eIF-4E phosphorylation is related to the defect in late viral mRNA export. However, recent data from HARADA and BERK (1999) indicate that E1B-55kDa might perform an additional function in host cell shut-off and stimulation of late viral mRNA translation. In their studies, replication of the E1B-55kDa mutant virus *dl*1520 in p53-negative H1299 cells was temperature-dependent. Similar to the cold-sensitive growth reported for other E1B-55kDa mutants in HeLa cells (Ho et al. 1982; WILLIAMS et al. 1986; LEPPARD and SHENK 1989; GOODRUM and ORNELLES 1998), *dl*1520 replication was severely reduced at 32°C compared to that of wild-type Ad5, whereas at 39°C mutant virus yield was restored to near wild-type levels. Interestingly, as opposed to HeLa cells, infection of H1299 cells at 32°C resulted only in a modest reduction of cytoplasmic accumulation of late viral mRNAs, while late viral protein synthesis was substantially decreased. Thus, in H1299 cells the defect in late viral mRNA export cannot account for the inability of *dl*1520 to enhance late viral mRNA translation and shut-off of host protein synthesis. It is, therefore, possible that E1B-55kDa has a more direct role in inactivating eIF-4E. Harada and Berk proposed that stimulation of viral and inhibition of cellular mRNA translation by E1B-55kDa may result from the dephosphorylation of eIF-4E (HARADA and BERK 1999). Consequently, the failure of E1B-55kDa mutants to inhibit eIF-4E phosphorylation prevents host cell shut-off and the preferential translation of major late mRNAs. Such a model could also explain the observed elevated temperature compensation of E1B-55kDa mutation since inhibition of eIF-4E phosphorylation also occurs during the heat shock response (LAMPHEAR and PANNIERS 1991). If eIF-4E is indeed inactivated at 39°C, major late viral mRNA translation would clearly benefit from the now limiting concentrations of eIF-4F, as in wild-type adenovirus-infected cells. In this context it would be interesting to see whether infection of synchronized cells during S phase compensates for host cell shut-off and stimulation of major late mRNA translation defects of E1B-55kDa mutants. Finally, it is interesting to note that the role of E1B-55kDa in mRNA transport parallels its role in viral mRNA translation. In both cases E1B-55kDa is required to promote export and translation of viral mRNAs while inhibiting the same processes for cellular mRNAs. It remains to be established whether these activities are connected or independent phenomena.

5 Involvement of E1B-55kDa and E4orf6 in Posttranslational Control

Evidence for a role of E1B-55kDa and E4orf6 in regulating viral gene expression at the posttranslational level originates from recent studies showing that both early proteins are required to counteract the E1A-induced stabilization of p53 in virus-infected and transformed cells by accelerating p53 proteolytic degradation (MOORE et al. 1996; NEVELS et al. 1997; QUERIDO et al. 1997; STEEGENGA et al. 1998; ROTH et al. 1998; NEVELS et al. 1999a; WIENZEK et al. 2000). Because E1B-55kDa and E4orf6 associate in virus-infected cells (see Sect. 2) and bind to different domains on p53 (LIN et al. 1994; DOBNER et al. 1996), it has been proposed that a trimeric complex is the active form that targets p53 for degradation (NEVELS et al. 1997; QUERIDO et al. 1997), although at present there is no direct evidence that such a complex exists. Destabilization of p53 in transformed rat cells is dependent on E1B-55kDa/E4orf6 complex formation and requires a highly conserved cysteine-rich sequence motif in the central part of E4orf6 (Fig. 4) that may connect the protein to cellular factors involved in ubiquitin-dependent proteolytic degradation pathways (NEVELS et al. 2000). Thus, the E1B-55kDa/E4orf6 proteins may provide a platform for the degradation of certain cellular proteins that bind to the E1B-55kDa/E4orf6 complex. This idea is supported by work from GRIFMAN et al. (1999), who investigated the requirements for efficient rAAV transduction mediated by the E4orf6 protein (SAMULSKI and SHENK 1988; HUANG and HEARING 1989; FERRARI et al. 1996; FISHER et al. 1996). In their study, enhanced viral transduction correlated with E4orf6-induced degradation of cyclin A accompanied by an accumulation of cells in S phase. As discussed (see Sect. 3.6), the region on the E4orf6 protein essential for helping rAAV maps to a putative RXL motif within the arginine-rich amphipathic α-helix and this probably also mediates binding to cyclin A and E1B-55kDa (GRIFMAN et al. 1999). Collectively, these observations indicate that E1B-55kDa and E4orf6 may modulate cell growth control and relieve growth restrictions imposed on viral replication by the cell cycle through, at least in part, proteolytic degradation of cell cycle regulatory proteins. Furthermore, they raise the intriguing possibility that this activity may also play a role in the E1B-55kDa/E4orf6-dependent stimulation of viral and inhibition of cellular mRNA export.

6 Conclusions and Future Directions

Over the past 10 years, work by a number of investigators has provided novel and important insights into the molecular mechanism by which adenovirus proteins control viral and cellular mRNA export. It is generally now accepted that regulated viral late mRNA transport involves the concerted action of three viral early proteins, E4orf3, E4orf6 and E1B-55kDa, which form different complexes in the nucleus of virus-infected cells. The observation that the E4orf3 and E4orf6 interactions with E1B-55kDa are probably mutually exclusive (LEPPARD and EVERETT 1999; KÖNIG et al. 1999) and are regulated in a temporally coordinated manner suggests that E4orf3/E1B-55kDa and E4orf6/E1B-55kDa protein complexes are diversely important for efficient late viral biogenesis. Another possible interpretation is that the overlapping functions of both E4 proteins in late viral gene expression relate, at least in part, to their complex formation with E1B-55kDa. The available data suggest that selective export of late viral mRNAs is linked to transcriptional activation during the late phase of infection (Sect. 3.5), is specified by the E1B-55kDa/E4orf6 interaction and is enhanced by E4orf3 and E4orf6, which may cotranscriptionally modulate splicing commitment factors with consequent effects on RNA stability and recruitment into export pathways (Sect. 3.4). One could predict that E1B-55kDa/E4orf6 posttranscriptionally counteracts nuclear retention of incompletely spliced viral late mRNAs sequestered in IG clusters (Sect. 3.3) in conjunction with E1B-AP5 (Sect. 3.5) and directly participate in nucleocytoplasmic RNA trafficking presumably through a Rev NES-dependent export pathway (Sect. 3.6).

To some extent, E1B-55kDa/E4orf6 operates in a similar fashion to HIV-1 Rev. This implies that adenovirus and complex retroviruses use, at some stage, the same or overlapping pathways for late viral mRNA export, which are possibly disconnected from general mRNA export pathways. In view of this similarity, one might speculate that a common step involved in adenovirus and HIV late mRNA export involves the interaction with CRM1. Presumably this includes movement to the nuclear face of the NPC, interactions between CRM1 and FG-containing nucleoporins (FARJOT et al. 1999), docking, and translocation through the NPC after Ran-GTP hydrolysis. Although such a model is attractive, it remains to be determined whether E1B-55kDa, E4orf6 and/or the viral protein complex shuttle in adenovirus-infected cells. The fact that both E1B-55kDa and E4orf6 contain leucine-rich NESs of the Rev-type is intriguing. Studies on Rev indicate that multiple CRM1 export receptor molecules need to be recruited to the Rev/RRE/RNPs for export to be activated (reviewed in POLLARD and MALIM 1998). We speculate that a similar situation may apply for the E1B-55kDa/E4orf6 NES-dependent export. Alternatively and/or additionally, the E1B-55kDa and E4orf6 NESs may be functionally compensatory. Such a mechanism could explain the finding that E4orf6 lacking a functional NES allows lytic growth of an E4-deleted mutant virus in transfection/infection experiments in HeLa cells (WEIGEL and DOBBELSTEIN 2000) and could reflect the need to infect many different types of cells.

Although many of the considerations discussed in this chapter are attractive and may reflect critical features of regulated mRNA export in adenovirus-infected cells, numerous critical aspects remain unsolved. Clearly further work is needed to solve the apparent contradictory results with regard to the ability of E4orf6 to shuttle in the presence (DOBBELSTEIN et al. 1997) or absence (GOODRUM et al. 1996; ORLANDO and ORNELLES 1999; FISCHER et al. 1999) of E1B-55kDa and the conflicting observation that the E4orf6 NES fused to GST or GFP does not mediate shuttling after microinjection into the nuclei of HeLa cells (KRÄTZER et al. 2000). Given the multiple functions of E1B-55kDa and E4orf6 in adenovirus-infected cells, it is almost certain that other cellular factors besides E1B-AP5 are targets of both these early proteins in the assembly, release and shuttling of viral RNPs. The question is what are these host cell factors? Also, is CRM1 the export receptor for both E1B-55kDa and E4orf6? Are different export pathways used by E4orf6 and E1B-55kDa NESs? And what downstream components mediate docking to and translocation through the NPC? Does the latter process involve Ran and Ran-GTP hydrolysis? Moreover, upon reaching the cytoplasm there are more questions concerning how the viral RNP/E1B/E4 complexes are disassembled and what the exact signals and factors that confer nuclear import of the viral proteins might be.

In addition, it still remains an open question as to whether the E1B-55kDa/E4orf6-dependent enhanced transport of viral late mRNAs and the inhibition of transport of cellular mRNAs are connected or independent phenomena, and why certain transcripts from activated cellular genes during the late phase of infection escape the transport block but require E1B-55kDa for efficient export (Sect. 3.5). One hypothesis, which unites many of the above considerations, is that E1B-55kDa/E4orf6-mediated RNA transport is divorced from general mRNA export pathways. Possibly 55kDa/E4orf6 usurps a Rev NES-dependent export pathway, presumably involving CRM1 as an export receptor, while simultaneously inhibiting cellular mRNA export either by recruiting essential cellular mRNA transport factors into the viral transcription/replication centers or by posttranslationally modulating the activity and/or stability of these components (Sect. 5). Indeed, since cellular messages that escape the virus-induced export block might represent a stress-specific response, by analogy to regulation of hs RNA export in *S. cerevisiae* (Sect. 3.6), their E1B-55kDa-dependent cytoplasmic transport could be explained if E1B-55kDa/E4orf6 dominates a CRM1-dependent "emergency" export pathway used under stress conditions.

Finally, we are still far from understanding the mechanisms underlying the differences in E1B-55kDa and E4orf6 mutant virus growth in various tumor and non-immortalized primary cell strains and how these are linked to regulated viral mRNA export (Sect. 3.7) and translation (Sect. 4). Obviously, these differences may be explained by the presence of cellular components that, in the case of non-restrictive cells, compensate for E1B-55kDa/E4orf6 late-phase functions. For example, the E4orf6-dependent localization of E1B-55kDa to the viral transcription/replication centers in restrictive HeLa cells (ORNELLES and SHENK 1991) is apparently compensated by some cellular activity in A549 cells (KÖNIG et al.

1999). This could explain, at least in part, why the effect of E4orf6 mutation on late viral gene expression is less severe in A549 cells than in HeLa cells (HALBERT et al. 1985; CUTT et al. 1987; SMILEY et al. 1990). In particular, knowledge of the molecular basis of these mechanisms could be directly applied to the use of oncolytic adenovirus vectors in human tumor therapy. Moreover, an important aspect, which has not been addressed in this chapter, is the observation that all three early proteins promote oncogenic transformation of non-permissive rodent cells in cooperation with E1A (BARKER and BERK 1987; MOORE et al. 1996; NEVELS et al. 1997, 1999b). This raises the obvious question of whether their lytic activities overlap with their roles in transformation. In particular, it will be interesting to see whether nucleocytoplasmic shuttling of E1B-55kDa and E4orf6 is an integral part by which both early proteins contribute to oncogenic transformation. All of these questions warrant continued investigation in the future and will perhaps reveal new principles of mRNA transport in general, which will certainly contribute to designing highly effective and safe adenovirus vectors for human gene therapy.

Acknowledgements. We are most grateful to Jane Flint, Michael Nevels, Birgitt Täuber for constructive comments on the manuscript, Elisa Izaurralde for providing unpublished information and Keith Leppard for the transcription and translation map of Ad2 on which Fig. 1 is based. Work in this laboratory was supported by grants from the Deutsche Forschungsgemeinschaft (DFG) to T.D.

References

Aspegren A, Rabino C, Bridge E (1998) Organization of splicing factors in adenovirus-infected cells reflects changes in gene expression during the early to late phase transition. Exp Cell Res 245:203–213
Babich A, Feldman LT, Nevins JR, Darnell JE, Weinberger C (1983) Effect of adenovirus on metabolism of specific host mRNAs: transport control and specific translation discrimination. Mol Cell Biol 3:1212–1221
Babiss LE, Ginsberg HS (1984) Adenovirus type 5 early region 1b gene product is required for efficient shutoff of host protein synthesis. J Virol 50:202–212
Babiss LE, Ginsberg HS, Darnell JJ (1985) Adenovirus E1B proteins are required for accumulation of late viral mRNA and for effects on cellular mRNA translation and transport. Mol Cell Biol 5:2552–2558
Bachi A, Braun IC, Rodrigues JP, Panté N, Ribbeck K, von Kobbe C, Kutay U, Wilm M, Görlich D, Carmo-Fonseca M, Izaurralde E (2000) The C-terminal domain of TAP interacts with the nuclear pore complex and promotes export of specific CTE-bearing RNA substrates. RNA 6:136–158
Barker DD, Berk AJ (1987) Adenovirus proteins from both E1B reading frames are required for transformation of rodent cells by viral infection and DNA transfection. Virology 156:107–121
Beltz GA, Flint SJ (1979) Inhibition of HeLa cell protein synthesis during adenovirus infection: restriction of cellular messenger RNA sequences to the nucleus. J Mol Biol 131:353–373
Benihoud K, Yeh P, Perricaudet M (1999) Adenovirus vectors for gene delivery. Curr Opin Biotech 10:440–447
Berkner KL, Sharp PA (1985) Effect of the tripartite leader on synthesis of a non-viral protein in an adenovirus 5 recombinant. Nucleic Acids Res 13:841–857
Bernards R, de Leeuw MG, Houweling A, van der Eb AJ (1986) Role of the adenovirus early region 1B tumor antigens in transformation and lytic infection. Virology 150:126–139
Bilbao G, Contreras JL, Gómez-Navarro J, Curiel DT (1998) Improving adenovirus vectors for gene therapy. Tumor Targeting 3:59–79

Bischoff JR, Kirn DH, Williams A, Heise C, Horn S, Muna M, Ng L, Nye JA, Sampson Johannes A, Fattaey A, McCormick F (1996) An adenovirus mutant that replicates selectively in p53-deficient human tumor cells. Science 274:373–376

Boyer J, Rohleder K, Ketner G (1999) Adenovirus E4 34k and E4 11k inhibit double strand break repair and are physically associated with the cellular DNA-dependent protein kinase. Virology 263:307–312

Bridge E, Ketner G (1989) Redundant control of adenovirus late gene expression by early region 4. J Virol 63:631–638

Bridge E, Ketner G (1990) Interaction of adenoviral E4 and E1b products in late gene expression. Virology 174:345–353

Bridge E, Carmo-Fonseca M, Lamond A, Petterson U (1993) Nuclear organization of splicing small nuclear ribonucleoproteins in adenovirus-infected cells. J Virol 76:5792–5802

Bridge E, Medghalchi S, Ubol S, Leesong M, Ketner G (1993) Adenovirus early region 4 and viral DNA synthesis. Virology 193:794–801

Bridge E, Pettersson U (1995) Nuclear organization of replication and gene expression in adenovirus-infected cells. Curr Top Microbiol Immunol 199:99–117

Bridge E, Pettersson U (1996) Nuclear organization of adenovirus RNA biogenesis. Exp Cell Res 229:233–239

Bridge E, Riedel KU, Johansson BM, Pettersson U (1996) Spliced exons of adenovirus late RNAs colocalize with snRNP in a specific nuclear domain. J Cell Biol 135:303–314

Carvalho T, Seeler JS, Ohman K, Jordan P, Pettersson U, Akusjärvi G, Carmo Fonseca M, Dejean A (1995) Targeting of adenovirus E1A and E4-ORF3 proteins to nuclear matrix-associated PML bodies. J Cell Biol 131:45–56

Chang DD, Sharp PA (1989) Regulation by HIV Rev depends upon recognition of splice sites. Cell 59:789–795

Cole CN, Hammell CM (1998) Nucleocytoplasmic transport: driving and directing transport. Curr Biol 8:368–372

Cullen BR (1998) Posttranscriptional regulation by the HIV-1 Rev protein. Semin Virol 8:327–334

Cutt JR, Shenk T, Hearing P (1987) Analysis of adenovirus early region 4-encoded polypeptides synthesized in productively infected cells. J Virol 61:543–552

Denome RM, Werner EA, Patterson RJ (1989) RNA metabolism in nuclei: adenovirus and heat shock alter intranuclear RNA compartmentalization. Nucleic Acids Res 17:2081–2098

Dix I, Leppard KN (1993) Regulated splicing of adenovirus type 5 E4 transcripts and regulated cytoplasmic accumulation of E4 mRNA. J Virol 67:3226–3231

Dobbelstein M, Roth J, Kimberly WT, Levine AJ, Shenk T (1997) Nuclear export of the E1B 55-kDa and E4 34-kDa adenoviral oncoproteins mediated by a rev-like signal sequence. EMBO J 16:4276–4284

Dobner T, Horikoshi N, Rubenwolf S, Shenk T (1996) Blockage by adenovirus E4orf6 of transcriptional activation by the p53 tumor suppressor. Science 272:1470–1473

Dolph PJ, Racaniello V, Villamarin A, Palladino F, Schneider RJ (1988) The adenovirus tripartite leader may eliminate the requirement for cap-binding protein complex during translation initiation. J Virol 62:2059–2066

Dolph PJ, Huang JT, Schneider RJ (1990) Translation by the adenovirus tripartite leader: elements which determine independence from cap-binding protein complex. J Virol 64:2669–2677

Doucas V, Ishov AM, Romo A, Juguilon H, Weitzman MD, Evans RM, Maul GG (1996) Adenovirus replication is coupled with the dynamic properties of the PML nuclear structure. Genes Dev 10:196–207

Fackelmayer FO, Dahm K, Renz A, Ramsperger U, Richter A (1994) Nucleic-acid-binding properties of hnRNP-U/SAF-A, a nuclear-matrix protein which binds DNA and RNA in vivo and in vitro. Eur J Biochem 221:749–757

Farjot G, Sergeant A, Mikaelian I (1999) A new nucleoporin-like protein interacts with both HIV-1 Rev nuclear export signal and CRM-1. J Biol Chem 274:17309–17317

Ferrari FK, Samulski T, Shenk T, Samulski RJ (1996) Second-strand synthesis is a rate-limiting step for efficient transduction by recombinant adeno-associated virus vectors. J Virol 70:3227–3234

Fischer N, Voss MD, Mueller-Lantzsch N, Grässer FA (1999) A potential NES of the Epstein-Barr virus nuclear antigen 1 (EBNA1) does not confer shuttling. FEBS Lett 447:311–314

Fisher KJ, Gao GP, Weitzman MD, DeMatteo R, Burda JF, Wilson JM (1996) Transduction with recombinant adeno-associated virus for gene therapy is limited by leading-strand synthesis. J Virol 70:520–532

Gabler S, Schütt H, Groitl P, Wolf H, Shenk T, Dobner T (1998) E1B 55-kilodalton-associated protein: a cellular protein with RNA-binding activity implicated in nucleocytoplasmic transport of adenovirus and cellular mRNAs. J Virol 72:7960–7971

Gaynor RB, Hillman D, Berk AJ (1984) Adenovirus early region 1A protein activates transcription of a novel gene introduced into mammalian cells by infection or transfection. Proc Natl Acad Sci USA 81:1193–1197

Goodrum FD, Shenk T, Ornelles DA (1996) Adenovirus early region 4 34-kilodalton protein directs the nuclear localization of the early region 1B 55-kilodalton protein in primate cells. J Virol 70:6323–6335

Goodrum FA, Ornelles DA (1997) The early region 1B 55-kilodalton oncoprotein of adenovirus relieves growth restrictions imposed on viral replication by the cell cycle. J Virol 71:548–561

Goodrum FD, Ornelles DA (1998) p53 status does not determine outcome of E1B 55-Kilodalton mutant adenovirus lytic infection. J Virol 72:9479–9490

Goodrum FD, Ornelles DA (1999) Roles for the E4 orf6, orf3, and E1B 55-kilodalton proteins in cell cycle-independent adenovirus replication. J Virol 73:7474–7488

Görlich D, Mattaj IW (1996) Nucleocytoplasmic transport. Science 271:1513–1518

Görlich D, Kutay U (1999) Transport between the cell nucleus and the cytoplasm. Annu Rev Cell Dev Biol 15:607–660

Grifman M, Chen NN, Gao G, Cathomen T, Wilson JM, Weitzman MD (1999) Overexpression of cyclin A inhibits augmentation of recombinant adeno-associated virus transduction by the adenovirus E4orf6 protein. J Virol 73:10010–10019

Halbert DN, Cutt JR, Shenk T (1985) Adenovirus early region 4 encodes functions required for efficient DNA replication, late gene expression, and host cell shutoff. J Virol 56:250–257

Hamm J, Mattaj IW (1990) Monomethylated cap structures facilitate RNA export from the nucleus. Cell 63:109–118

Harada JN, Berk AJ (1999) p53-independent and -dependent requirements for E1B-55k in adenovirus type 5 replication. J Virol 73:5333–5344

Harrison T, Graham F, Williams J (1977) Host-range mutants of adenovirus type 5 defective for growth in HeLa cells. Virology 77:319–329

Hearing P, Shenk T (1985) Sequence-independent auto-regulation of the adenovirus type E1A transcription unit. Mol Cell Biol 5:3214–3221

Heise C, Sampson Johannes A, Williams A, McCormick F, Von Hoff DD, Kirn DH (1997) ONYX-015, an E1B gene-attenuated adenovirus, causes tumor-specific cytolysis and antitumoral efficacy that can be augmented by standard chemotherapeutic agents. Nat Med 3:639–645

Ho YS, Galos R, Williams J (1982) Isolation of type 5 adenovirus mutants with a cold-sensitive host range phenotype: genetic evidence of an adenovirus transformation maintenance function. Virology 122:109–124

Horridge JJ, Leppard KN (1998) RNA-binding activity of the E1B 55-kilodalton protein from human adenovirus type 5. J Virol 72:9374–9379

Horwitz MS (1996) Adenoviruses. In: Fields BN, Knipe DM, Howley PM (eds) Virology. Lippincott-Raven, New York, pp 2149–2171

Huang MM, Hearing P (1989) Adenovirus early region 4 encodes two gene products with redundant effects in lytic infection. J Virol 63:2605–2615

Huang JT, Schneider RJ (1991) Adenovirus inhibition of cellular protein synthesis involves inactivation of cap-binding protein. Cell 65:271–280

Huang W, Flint SJ (1998) The tripartite leader sequence of subgroup C adenovirus major late mRNAs can increase the efficiency of mRNA export. J Virol 72:225–235

Imperiale MJ, Akusjärvi G, Leppard KN (1995) Post-transcriptional control of adenovirus gene expression. Curr Top Microbiol Immunol 199:139–171

Ishov AM, Maul GG (1996) The periphery of nuclear domain 10 (ND10) as site of DNA virus deposition. J Cell Biol 134:815–826

Jiménez-Garcia LF, Spector DL (1993) In vivo evidence that transcription and splicing are coordinated by a recruiting mechanism. Cell 73:47–59

Kanopka A, Muhlemann O, Petersen Mahrt S, Estmer C, Ohrmalm C, Akusjärvi G (1998) Regulation of adenovirus alternative RNA splicing by dephosphorylation of SR proteins. Nature 393:185–187

Kiledjian M, Dreyfuss G (1992) Primary structure and binding activity of the hnRNP U protein: binding RNA through RGG box. EMBO J 11:2655–2664

Kleinberger T, Shenk T (1993) Adenovirus E4orf4 protein binds to protein phosphatase 2A, and the complex down regulates E1A-enhanced junB transcription. J Virol 67:7556–7560

König C, Roth J, Dobbelstein M (1999) Adenovirus type 5 E4orf3 protein relieves p53 inhibition by E1B-55-kilodalton protein. J Virol 73:2253–2262

Krätzer F, Rosorius O, Heger P, Hirschmann N, Dobner T, Hauber J, Stauber RH (2000) The adenovirus type 5 E1B-55k oncoprotein is a highly active shuttle protein and shuttling is independent of E4orf6, p53 and Mdm2. Oncogene 19:850–857

Lamphear BJ, Panniers R (1991) Heat shock impairs the interaction of cap-binding protein complex with 5′ mRNA cap. J Biol Chem 266:2789–2794

Lavoie JN, Nguyen M, Marcellus RC, Branton PE, Shore GC (1998) E4orf4, a novel adenovirus death factor that induces p53-independent apoptosis by a pathway that is not inhibited by zVAD-fmk. J Cell Biol 140:637–645

Legrain P, Rosbash M (1989) Somce cis- and trans-acting mutants for splicing target pre-mRNA to the cytoplasm. Cell 57:573–583

Leppard KN, Shenk T (1989) The adenovirus E1B 55 kd protein influences mRNA transport via an intranuclear effect on RNA metabolism. EMBO J 8:2329–2336

Leppard KN (1993) Selective effects on adenovirus late gene expression of deleting the E1b 55K protein. J Gen Biol 74:575–582

Leppard KN (1997) E4 gene function in adenovirus, adenovirus vector and adeno-associated virus infections. J Gen Biol 78:2131–2138

Leppard KN (1998) Regulated RNA processing and RNA transport during adenovirus infection. Semin Virol 8:301–307

Leppard KN, Everett RD (1999) The adenovirus type 5 E1b 55K and E4 Orf3 proteins associate in infected cells and affect ND10 components. J Gen Biol 80:997–1008

Liang S, Hitomi M, Tartakoff AM (1995) Adenoviral E1B-55kDa protein inhibits yeast mRNA export and perturbs nuclear structure. Proc Natl Acad Sci USA 92:7372–7375

Liao D, Yu A, Weiner AM (1999) Coexpression of the adenovirus 12 E1B 55kDa oncoprotein and cellular tumor suppressor p53 is sufficient to induce metaphase fragility of the human RNU2 locus. Virology 254:11–23

Lin J, Chen J, Elenbaas B, Levine AJ (1994) Several hydrophobic amino acids in the p53 amino-terminal domain are required for transcriptional activation, binding to mdm-2 and the adenovirus 5 E1B 55-kD protein. Genes Dev 8:1235–1246

Logan J, Shenk T (1984) Adenovirus tripartite leader sequence enhances translation of mRNAs late after infection. Proc Natl Acad Sci USA 81:3655–3659

Mathews MB, Shenk T (1991) Adenovirus virus-associated RNA and translation control. J Virol 65:5657–5662

Mattaj IW, Englmeier L (1998) Nucleocytoplasmic transport: the soluble phase. Annu Rev Biochem 67:265–306

Maul GG (1998) Nuclear domain 10, the site of DNA virus transcription and replication. BioEssays 20(8):660–667

Moore M, Schaack J, Baim SB, Morimoto RI, Shenk T (1987) Induced heat shock mRNAs escape the nucleocytoplasmic transport block in adenovirus-infected HeLa cells. Mol Cell Biol 7: 4505–4512

Moore MA, Shenk T (1988) The adenovirus tripartite leader sequence can alter nuclear and cytoplasmic metabolism of a non-adenovirus mRNA within infected cells. Nucleic Acids Res 16:2247–2262

Moore M, Horikoshi N, Shenk T (1996) Oncogenic potential of the adenovirus E4orf6 protein. Proc Natl Acad Sci USA 93:11295–11301

Müller U, Kleinberger T, Shenk T (1992) Adenovirus E4orf4 protein reduces phosphorylation of c-Fos and E1A proteins while simultaneously reducing the level of AP-1. J Virol 66:5867–5878

Nakielny S, Dreyfuss G (1999) Transport of proteins and RNAs in and out of the nucleus. Cell 99:677–690

Nevels M, Rubenwolf S, Spruss T, Wolf H, Dobner T (1997) The adenovirus E4orf6 protein can promote E1A/E1B-induced focus formation by interfering with p53 tumor suppressor function. Proc Natl Acad Sci USA 94:1206–1211

Nevels M, Spruss T, Wolf H, Dobner T (1999a) The adenovirus E4orf6 protein contributes to malignant transformation by antagonizing E1A-induced accumulation of the tumor suppressor protein p53. Oncogene 18:9–17

Nevels M, Täuber B, Kremmer E, Spruss T, Wolf H, Dobner T (1999b) Transforming potential of the adenovirus type 5 E4orf3 protein. J Virol 73:1591–1600

Nevels M, Rubenwolf S, Spruss T, Wolf H, Dobner T (2000) Two distinct activities contribute to the oncogenic potential of the adenovirus type 5 E4orf6 protein. J Virol (in press)

Neville M, Stutz F, Lee L, Davis LI, Rosbash M (1997) The importin-beta family member Crm1p bridges the interaction between Rev and the nuclear pore complex during nuclear export. Curr Biol 7:767–775

Nicolás AL, Munz PL, Falck-Pedersen E, Young CSH (2000) Creation and repair of specific DNA double-strand breaks in vivo following infection with adenovirus vectors expressing *Saccharomyces cervisiae* HO endonuclease. Virology 266:211–224

Nordqvist K, Akusjärvi G (1990) Adenovirus early region 4 stimulates mRNA accumulation via 5′ introns. Proc Natl Acad Sci USA 87:9543–9547

Nordqvist K, Öhman K, Akusjärvi G (1994) Human adenovirus encodes two proteins which have opposite effects on accumulation of alternatively spliced mRNAs. Mol Cell Biol 14:437–445

Öhman K, Nordqvist K, Akusjärvi G (1993) Two adenovirus proteins with redundant activities in virus growth facilitates tripartite leader mRNA accumulation. Virology 194:50–58

Orlando JS, Ornelles DA (1999) An arginine-faced amphipathic alpha helix is required for adenovirus type 5 E4orf6 protein function. J Virol 73:4600–4610

Ornelles DA, Shenk T (1991) Localization of the adenovirus early region 1B 55-kilodalton protein during lytic infection: association with nuclear viral inclusions requires the early region 4 34-kilodalton protein. J Virol 65:424–429

Pilder S, Moore M, Logan J, Shenk T (1986) The adenovirus E1B-55K transforming polypeptide modulates transport or cytoplasmic stabilization of viral and host cell mRNAs. Mol Cell Biol 6: 470–476

Pollard VW, Malim MH (1998) The HIV-1 rev protein. Annu Rev Microbiol 52:491–532

Pombo A, Ferreira J, Bridge E, Carmo-Fonseca M (1994) Adenovirus replication and transcription sites are spatially separated in the nucleus of infected cells. EMBO J 13:5075–5085

Puvion Dutilleul F, Pichard E (1992) Segregation of viral double-stranded and single-stranded DNA molecules in nuclei of adenovirus infected cells as revealed by electron microscope in situ hybridization. Biol Cell 76:139–150

Puvion-Dutilleul F, Chelbi-Alix MK, Koken M, Quignon F, Puvion E, de The H (1995) Adenovirus infection induces rearrangements in the intranuclear distribution of the nuclear body-associated PML protein. Exp Cell Res 218:9–16

Querido E, Marcellus R, Lai A, Rachel C, Teodoro JG, Ketner G, Branton PE (1997) Regulation of p53 levels by the E1B 55-kilodalton protein and E4orf6 in adenovirus-infected cells. J Virol 71:3788–3798

Rebelo L, Almeida F, Ramos C, Bohmann K, Lamond AI, Carmo Fonseca M (1996) The dynamics of coiled bodies in the nucleus of adenovirus-infected cells. Mol Biol Cell 7:1137–1151

Roth J, König C, Wienzek S, Weigel S, Ristea S, Dobbelstein M (1998) Inactivation of p53 but not p73 by adenovirus type 5 E1B 55-Kilodalton and E4 34-Kilodalton oncoproteins. J Virol 72:8510–8516

Rothmann T, Hengstermann A, Whitaker NJ, Scheffner M, zur Hausen H (1998) Replication of ONYX-015, a potential anticancer adenovirus, is independent of p53 status in tumor cells. J Virol 72:9470–9478

Roulston A, Marcellus RC, Branton PE (1999) Viruses and apoptosis. Annu Rev Microbiol 53:577–628

Rubenwolf S, Schütt H, Nevels M, Wolf H, Dobner T (1997) Structural analysis of the adenovirus type 5 E1B 55-kilodalton-E4orf6 protein complex. J Virol 71:1115–1123

Saavedra CA, Hammell CM, Heath CV, Cole CN (1997) Yeast heat shock mRNAs are exported through a distinct pathway defined by Rip1p. Genes Dev 11:2845–2856

Samulski RJ, Shenk T (1988) Adenovirus E1B 55-Mr polypeptide facilitates timely cytoplasmic accumulation of adeno-associated virus mRNAs. J Virol 62:206–210

Sandler AB, Ketner G (1989) Adenovirus early region 4 is essential for normal stability of late nuclear RNAs. J Virol 63:624–630

Sarnow P, Sullivan CA, Levine AJ (1982a) A monoclonal antibody detecting the adenovirus type 5-E1b-58Kd tumor antigen: characterization of the E1b-58Kd tumor antigen in adenovirus-infected and -transformed cells. Virology 120:510–517

Sarnow P, Ho YS, Williams J, Levine AJ (1982b) Adenovirus E1b-58kd tumor antigen and SV40 large tumor antigen are physically associated with the same 54kd cellular protein in transformed cells. Cell 28:387–394

Sarnow P, Hearing P, Anderson CW, Halbert DN, Shenk T, Levine AJ (1984) Adenovirus early region 1B 58,000-dalton tumor antigen is physically associated with an early region 4 25,000-dalton protein in productively infected cells. J Virol 49:692–700

Seeler J-S, Dejean A (1999) The PML nuclear bodies: actors or extras? Curr Opin Gen Dev 9:362–367
Shenk T (1996) Adenoviridae: the viruses and their replication. In: Fields BN, Knipe DM, Howley PM (eds) Virology. Lippincott-Raven, New York, pp 2111–2148
Shtrichman R, Kleinberger T (1998) Adenovirus type 5 E4 open reading frame 4 protein induces apoptosis in transformed cells. J Virol 72:2975–2982
Shtrichman R, Sharf R, Barr H, Dobner T, Kleinberger T (1999) Induction of apoptosis by adenovirus E4-open-reading-frame-4 protein is specific to transformed cells and requires an interaction with protein phosphatase 2A. Proc Natl Acad Sci USA 96:10080–10085
Smiley JK, Young MA, Flint SJ (1990) Intranuclear location of the adenovirus type 5 E1B 55-kilodalton protein. J Virol 64:4558–4564
Steegenga WT, Riteco N, Jochemsen AG, Fallaux FJ, Bos JL (1998) The large E1B protein together with the E4orf6 protein target p53 for active degradation in adenovirus infected cells. Oncogene 16:349–357
Sternsdorf T, Grotzinger T, Jensen K, Will H (1997) Nuclear dots: actors on many stages. Immunobiology 198:307–331
Stutz F, Kantor J, Zhang D, McCarthy T, Neville M, Rosbash M (1997) The yeast nucleoporin rip1p contributes to multiple export pathways with no essential role for its FG-repeat region. Genes Dev 11:2857–2868
Stutz F, Neville M, Rosbash M (1995) Identification of a novel nuclear pore-associated protein as a functional target of the HIV-1 Rev protein in yeast. Cell 82:495–506
Stutz F, Rosbash M (1998) Nuclear RNA export. Genes Dev 12:3303–3310
Turnell AS, Grand RJA, Gallimore PH (1999) The replicative capacities of large E1B-null group A and group C adenoviruses are independent of host cell p53 status. J Virol 73:2074–2083
von Kries JP, Buck F, Stratling WH (1994) Chicken MAR binding protein p120 is identical to human heterogeneous nuclear ribonucleoprotein (hnRNP) U. Nucleic Acids Res 22:1215–1220
Weiden MD, Ginsberg HS (1994) Deletion of the E4 region of the genome produces adenovirus DNA concatemers. Proc Natl Acad Sci USA 91:153–157
Weigel S, Dobbelstein M (2000) The nuclear export signal within the E4orf6 protein of adenovirus type 5 supports virus replication and cytoplasmic accumulation of viral mRNA. J Virol 74:764–772
Weinberg DH, Ketner G (1986) Adenoviral early region 4 is required for efficient viral DNA replication and for late gene expression. J Virol 57:833–838
White E (1998) Regulation of apoptosis by adenovirus E1A and E1B oncoproteins. Semin Virol 8:505–513
Wienzek S, Roth J, Dobbelstein M (2000) E1B 55-kilodalton oncoproteins of adenovirus types 5 and 12 inactivate and relocalize p53, but not p51 or p73, and cooperate with E4orf6 proteins to destabilize p53. J Virol 74:193–202
Williams J, Karger BD, Ho YS, Castiglia CL, Mann T, Flint SJ (1986) The adenovirus E1B 495R protein plays a role in regulating the transport and stability of the viral late messages. Cancer Cells 4:275–284
Williams RD, Leppard KN (1996) Human immunodeficiency virus type 1 Rev-dependent effects on the late gene expression of recombinant human adenovirus. Virus Genes 13:111–120
Yang UC, Huang W, Flint SJ (1996) mRNA export correlates with activation of transcription in human subgroup C adenovirus-infected cells. J Virol 70:4071–4080
Yew PR, Kao CC, Berk AJ (1990) Dissection of functional domains in the adenovirus 2 early 1B 55K polypeptide by suppressor-linker insertional mutagenesis. Virology 179:795–805
Yew PR, Berk AJ (1992) Inhibition of p53 transactivation required for transformation by adenovirus early 1B protein. Nature 357:82–85
Zhang Y, Schneider RT (1993) Adenovirus inhibition of cellular protein synthesis and the specific translation of late viral mRNAs. Semin Virol 4:229–236
Zhang Y, Feigenbaum D, Schneider RJ (1994) A late adenovirus factor induces eIF-4E dephosphorylation and inhibition of cell protein synthesis. J Virol 68:7040–7050

Nuclear Export Mediated by the Rev/Rex Class of Retroviral *Trans*-activator Proteins

J. HAUBER

1 Introduction.	55
2 Rev and Rex Are Specific RNA-Binding Proteins	56
3 Functional Regions in Rev and Rex	58
4 Cellular Factors Interacting with Specific Domains in Rev and Rex	61
5 Perspectives	67
References	69

1 Introduction

Replication of human retroviruses, such as the human immunodeficiency virus type 1 (HIV-1) and the human T-cell leukemia virus type 1 (HTLV-1), depends on the transport of subgenomic mRNAs and unspliced genomic RNA from the nucleus to the cytoplasm. This nuclear RNA export is promoted by two small viral regulatory proteins, termed Rev in HIV-1 and Rex in HTLV-1. Both Rev and Rex interact with multiple cellular proteins in order to translocate viral RNA across the nuclear envelope and are considered to be model systems in which to study the regulation of nuclear export. In particular, Rev's function in nuclear export has been discussed as part of multiple general reviews on nucleocytoplasmic trafficking (STUTZ and ROSBASH 1998; IZAURRALDE and ADAM 1998; NAKIELNY and DREYFUSS 1999; GÖRLICH and KUTAY 1999). Moreover, an excellent and comprehensive review on the various aspects of HIV-1 Rev *trans*-activation has already been published (POLLARD and MALIM 1998). Instead of reiterating these data, in this chapter I would like to focus mainly on the various interactions of Rev and Rex with potential cellular cofactors. In addition, I would also like to explore some unresolved questions with respect to Rev/Rex function.

Institute for Clinical and Molecular Virology, University of Erlangen-Nürnberg, Schlossgarten 4, 91054 Erlangen, Germany

2 Rev and Rex Are Specific RNA-Binding Proteins

The genomes of human retroviruses are relatively small (~10kb) and are characterized by a complex organization (Fig. 1). Like every retrovirus, they contain genes encoding the structural proteins and enzymes Gag, Pol and Env. In addition, these genomes also encode multiple regulatory and accessory functions. In order to overcome the limitations resulting from the relatively small genome size, the genetic information has to be arranged by using overlapping genes and different reading frames. The transcriptional activation of the proviral 5'-long terminal repeat promoter (5'-LTR) results in the synthesis of the primary viral transcript that terminates in the 3'-LTR. Since multiple genes have to be expressed, this full-length RNA is subjected to alternative splicing in the nucleus in order to produce the various viral mRNA species that are subsequently exported to the cytoplasm.

In the case of HIV-1, three classes of RNAs have been observed (Fig. 1). The 2-kb class is multiply spliced, thereby deleting *gag-pol* and *env* sequences, and encodes the regulatory proteins Tat, Nef or Rev. The 4-kb class of viral mRNAs, which are used for the synthesis of Env or the accessory proteins Vif, Vpr or Vpu, is characterized by a gag-pol deletion resulting from a single splicing event. Finally, the 9-kb class consists of unspliced primary transcripts that are used either for Gag and Pol expression or act as genomes in the formation of progeny viruses. Processing of full-length viral RNA by the host splicing machinery in the nucleus is known to be inefficient compared to the splicing of cellular transcripts (CHANG and SHARP 1989). Thus, unspliced or incompletely spliced viral mRNAs are characterized by a relatively long nuclear half-life (MALIM and CULLEN 1993). This extended half-life depends on the presence of various sequence elements in the respective viral RNAs. It has been shown that viral splice sites are very inefficiently recognized by the nuclear host cell splicing machinery (CHANG and SHARP 1989; HAMMARSKJÖLD et al. 1994; STAFFA and COCHRANE 1994). In addition, the unspliced and incompletely spliced viral mRNAs contain sequence elements, termed either *cis*-acting repressor sequences (CRSs) or inhibitory sequences (INSs), which inhibit expression in the absence of Rev (COCHRANE et al. 1991; MALDARELLI et al. 1991; SCHWARTZ et al. 1992a,b). Thus, the combination of suboptimal splice sites plus inhibitory sequence elements cause a high level of intron-containing viral RNAs in the nucleus. These transcripts are eventually degraded or, upon expression of Rev, exported to the cytoplasm (MALIM and CULLEN 1993). The target site of Rev on these RNAs is a complex stem-loop structure, termed the Rev response element (RRE). The RRE is part of the *env* intron (MALIM et al. 1989a; HADZOPOULOU-CLADARAS et al. 1989; ROSEN et al. 1988) and contains a single, primary, high-affinity Rev-binding site and possibly multiple, secondary low-affinity sites for Rev interaction (DAEFLER et al. 1990; DALY et al. 1989; HEAPHY et al. 1990; MALIM et al. 1990; OLSEN et al. 1990a; ZAPP and GREEN 1989; COCHRANE et al. 1990a; for a detailed discussion of the exact sequence features of the RRE see POLLARD and MALIM 1998). Therefore, all

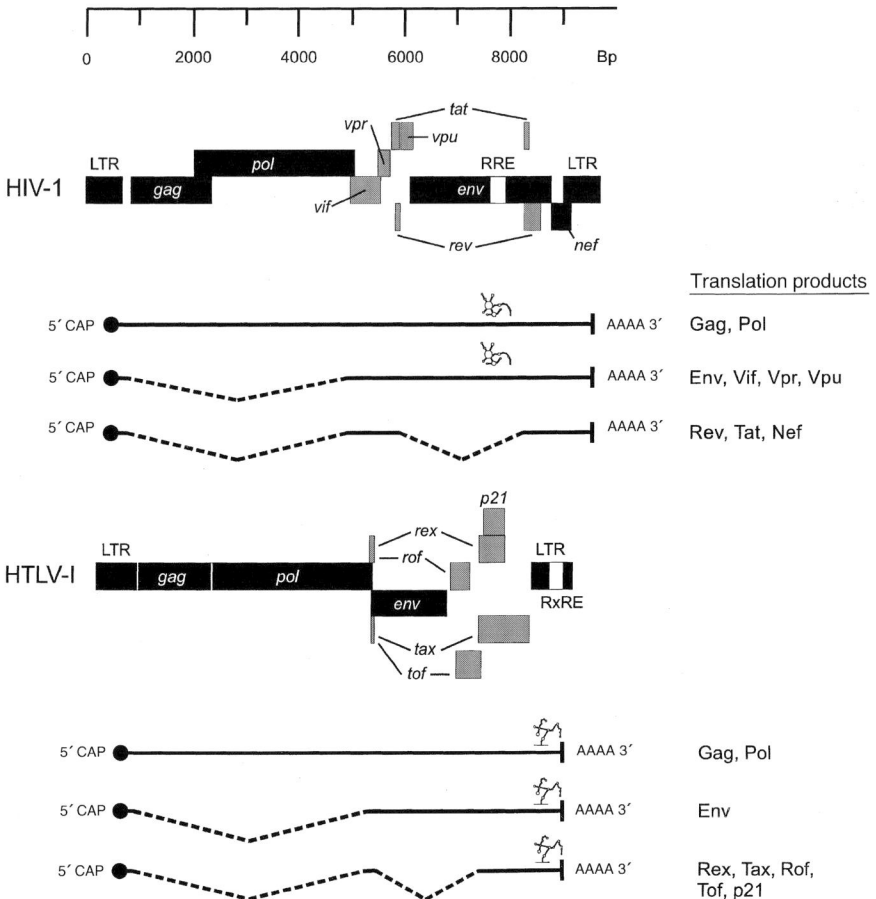

Fig. 1. Genomic organization of the proviral forms of the human immunodeficiency virus type 1 (HIV-1) and human T-cell leukemia virus type 1 (HTLV-1). The proviral genomes are flanked by long terminal repeats (*LTR*s). The structural proteins and enzymes are encoded by *gag*, *pol* and *env*. In addition, both genomes encode multiple regulatory and accessory gene products, particularly Rev, Tat, Nef, Vif, Vpr and Vpu in HIV-1 and Rex, Tax, Rof and Tof in HTLV-1. Gene expression results in three classes of mRNAs, which are shown below the respective proviral genomes. The unspliced viral mRNA serves as the genome in the production of the respective progeny viruses and also encodes Gag and Pol. Singly spliced transcripts are used for the protein synthesis of Env. In HIV-1, these messages also encode the accessory gene products Vif, Vpr and Vpu. In HIV-1, fully spliced (double-spliced) mRNAs encode the regulatory functions Rev, Tat and Nef and in HTLV-1 Rex, Tax, Rof, Tof and the Rex splice-variant p21. Rev and Rex are essential for virus replication and mediate nuclear export of unspliced and single-spliced mRNAs in the respective viral life cycles. The Rev response element (RRE), the RNA target sequence of Rev, is encoded by *env* sequences and is therefore only part of unspliced and single-spliced HIV-1 mRNAs. The RNA target sequence of Rex, the Rex response element (RxRE), is encoded by sequences of the 3′-LTR and is therefore an integral part of all HTLV-1 mRNAs. As indicated, both *cis*-acting RNA elements are characterized by a typical secondary stem-loop structure, which in each case generates a primary high-affinity binding site and multiple secondary interaction sites for Rev and Rex, respectively. Also note that Rev only binds its homologous RRE, while Rex is able to interact with both the homologous RxRE and the heterologous RRE

incompletely spliced and unspliced viral RNAs that encode the viral structural proteins and enzymes are subject to Rev regulation. In contrast, fully spliced mRNAs, which encode the regulatory proteins including Rev, are expressed in a Rev-independent manner.

The RNA target sequence of the HTLV-1 Rex *trans*-activator is also a highly structured RNA element, termed the Rex response element (RxRE). The RxRE is encoded by sequences within the 3'-LTR, making it an integral part of all viral mRNAs (SEIKI et al. 1988; TOYOSHIMA et al. 1990; HANLY et al. 1989; AHMED et al. 1990; BASKERVILLE et al. 1995). In addition to its role in mediating Rex responsiveness, the RxRE is also involved in the 3' processing of viral primary transcripts. The viral polyadenylation signal in HTLV-1 is separated from the 3' cleavage site by the RxRE sequence, a distance that does not allow processing of the 3' ends of the viral primary transcripts. Formation of the correct RxRE secondary structure, however, brings the two elements in close proximity to each other, thereby permitting correct polyadenylation (AHMED et al. 1991). In addition to directly binding the RxRE (BOGERD et al. 1991; BALLAUN et al. 1991; UNGE et al. 1991), Rex has also been shown to interact with the HIV-1 RRE and is able to functionally substitute for Rev in HIV-1 (RIMSKY et al. 1988). Although the biological activity of Rex on the heterologous RRE is clearly lower than the activity of the homologous Rev protein, this finding nevertheless indicates that Rev and Rex might exploit identical cellular cofactors in order to transport viral RNA across the nuclear envelope.

3 Functional Regions in Rev and Rex

Multiple studies in which mutational analysis was used to delineate functionally important regions in Rev and Rex have demonstrated that both retroviral *trans*-activators are characterized by a pronounced modular domain organization (Fig. 2). In both proteins, nuclear import and RNA-binding is mediated by a short stretch of amino acids rich in basic residues (COCHRANE et al. 1990b; HAMMERSCHMID et al. 1994; HOPE et al. 1990; KJEMS et al. 1992; ZAPP et al. 1991; VENKATESH et al. 1990; BOGERD et al. 1991; SIOMI et al. 1988; BÖHNLEIN et al. 1991a; KUBOTA et al. 1991; NOSAKA et al. 1989). Therefore, these regions are considered to be the nuclear localization signal (NLS) RNA-binding domain (RBD). This domain is located at the very amino-terminus of the 189-amino-acid Rex protein (amino acids 1–19), while the functionally corresponding region in the 116-amino-acid Rev protein spans residues 33–46. At steady-state, both proteins appear to localize in the nucleus, particularly in the nucleoli of expressing cells (CULLEN et al. 1988; COCHRANE et al. 1990b; SIOMI et al. 1988; NOSAKA et al. 1989, 1995). The identification of dominant-negative (*trans*-dominant) Rev/Rex mutants, which retain wild-type RNA-binding and nuclear localization characteristics, indicated the presence of an effector or activation domain in both proteins (MALIM et al. 1989b,

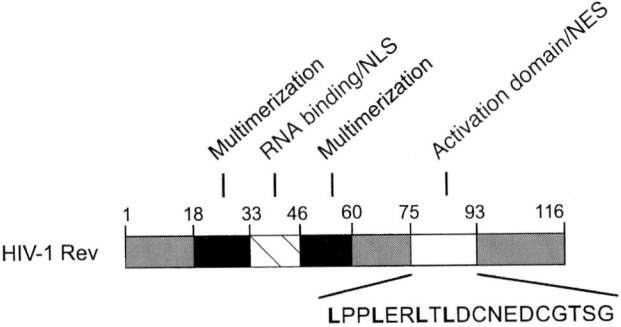

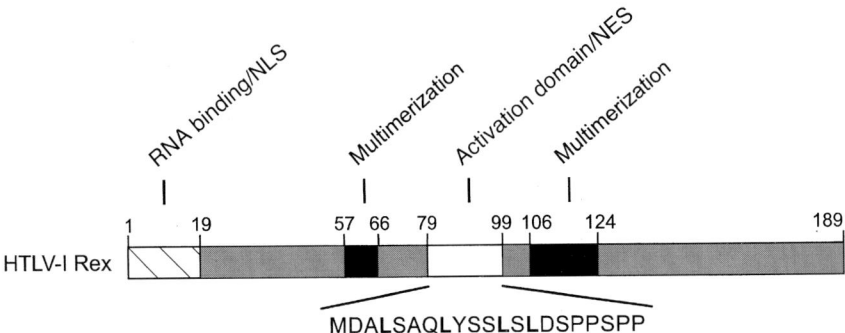

Fig. 2. The HIV-1 Rev and HTLV-1 Rex *trans*-activator proteins are characterized by a distinct domain organization. Nuclear import and interaction with the respective RNA target sequences (RRE in HIV-1; RxRE in HTLV-1) is mediated by short stretches of amino acids rich in basic residues. This RNA-binding/NLS domain localizes to amino acids 1–19 in the 189-amino-acid HTLV-1 Rex protein and to amino acids 33–46 in the 116-amino-acid HIV-1 Rev protein (*hatched boxes*). Regions involved in protein homo-multimer formation map to amino acids 18–32 and 47–60 in Rev and to amino acids 57–66 and 106–124 in Rex (*filled boxes*). The protein activation domains, which contain a leucine-rich peptide core motif that serves as a NES, maps to amino acids 75–93 in Rev and 79–99 in Rex (*open boxes*). These activation domains have been shown to be functionally interchangeable between Rev and Rex, and their amino acid sequences are shown as expanded sections. Leucine residues critical for NES function are shown in *boldface*. *NLS* Nuclear localization signal, *NES* nuclear export signal

1991; RIMSKY et al. 1989; BÖHNLEIN et al. 1991b; MERMER et al. 1990). The critical feature of these regions (Rev amino acids 75–93, Rex amino acids 79–99) is a series of four evenly spaced leucine residues that create a nuclear export signal (NES) (HOPE et al. 1991; VENKATESH and CHINNADURAI 1990; WEICHSELBRAUN et al. 1992; WOLFF et al. 1995; FISCHER et al. 1995; WEN et al. 1995; MEYER et al. 1996; KIM et al. 1996; BOGERD et al. 1996). The HIV-1 Rev protein was the first protein in which this type of leucine-rich NES was described (FISCHER et al. 1995). Since Rev and Rex contain sequences that mediate both nuclear import (NLS) and nuclear export (NES), the viral regulators constantly shuttle between the nuclear and cytoplasmic compartments of infected cells (MEYER and MALIM 1994; KALLAND et al. 1994; PALMERI and MALIM 1996; STAUBER et al. 1995).

Another important and often overlooked feature of Rev/Rex with respect to *trans*-activation is the ability of both proteins to form homo-multimers. In particular, Rev appears to be an intrinsically sticky protein that easily forms multimers, even in solution (OLSEN et al. 1990b; BOGERD and GREENE 1993; NALIN et al. 1990; WINGFIELD et al. 1991). However, various studies using recombinant Rev and in vitro-transcribed RRE RNA have demonstrated that Rev binds as a monomer to its high-affinity RRE-binding site (SLIIB). Subsequent protein–protein interactions and the occupation of secondary binding sites then result in the binding of multimeric Rev complexes to the RRE (COLE et al. 1993; DALY et al. 1989, 1993a; COOK et al. 1991; MALIM and CULLEN 1991; Powell et al. 1995; HEAPHY et al. 1990; IWAI et al. 1992; MANN et al. 1994; TILEY et al. 1992; ZEMMEL et al. 1996; ZAPP and GREEN 1989). Functional analysis of multimerization-deficient Rev mutant proteins has demonstrated that multimer formation is indeed required for Rev function (MALIM and CULLEN 1991; OLSEN et al. 1990b; ZAPP et al. 1991; MADORE et al. 1994; THOMAS et al. 1998). The residues that participate in Rev multimerization extend away from the RNA-binding/NLS domain and include the amino acid residues at position 18–32 and 47–60 (MALIM and CULLEN 1991; MADORE et al. 1994; THOMAS et al. 1998; BRICE et al. 1999). Structural analyses have also allowed the development of a refined model of the Rev amino-terminus indicating that, although the Rev multimerization interface is generated by two separate protein regions, it may actually form a single hydrophobic interaction area in the functional protein structure (AUER et al. 1994; THOMAS et al. 1997, 1998).

In contrast to Rev, relatively little is known about the sequence requirements for Rex multimer formation. Deletion and substitution mutagenesis studies have revealed that apparently two regions in HTLV-1 Rex, localizing to amino acids 57–66 and 106–124, may be involved in the formation of protein homo-multimers (HEGER et al. 1998). Mutation of residues mapping to these regions has been shown to abrogate *trans*-activation activity and to prevent dimer formation in a mammalian two-hybrid system (BOGERD and GREENE 1993; BÖHNLEIN et al. 1991b; RIMSKY et al. 1989). Moreover, biological activity could be reconstituted in some of these inactive mutants by fusion to a heterologous GCN4-derived leucine-zipper dimerization interface (HEGER et al. 1998). Since the Rex protein appears to be difficult to express in bacteria, only limited in vitro RNA-binding data, such as RNA mobility shift experiments, are available. Although in some experiments the occurrence of distinct signals were detected upon addition of Rex protein to radiolabeled RxRE RNA, other studies have failed to resolve multiple, distinct protein–nucleic acid complexes (BOGERD et al. 1991; BALLAUN et al. 1991; GRÖNE et al. 1994; UNGE et al. 1991; ASKJAER and KJEMS 1998). Nevertheless, the combined in vitro and in vivo data available suggest that oligomerization of Rex is indeed required for its full biological activity. It is also important to note that, in contrast to Rev, the majority of *trans*-dominant Rex mutants that have been described to date have mutations localized in both of the proposed Rex multimerization domains (BÖHNLEIN et al. 1991b; RIMSKY et al. 1989).

4 Cellular Factors Interacting with Specific Domains in Rev and Rex

Shuttling factors have to traverse the nuclear envelope in both directions. This transport into and out of the nucleus is mediated by nuclear pore complexes (NPCs), which are integral parts of the nuclear envelope (see chapter by Fahrenkrog et al., this volume). Although macromolecules of a molecular mass up to ~40kDa are thought to be able to diffuse freely through the aqueous pore channels, even small transport cargoes (<40kDa) are often translocated through the NPC in a signal-dependent manner. This directional transport is mediated by specific nuclear import and nuclear export receptors. As outlined before, Rev and Rex contain distinct regions that act as signal sequences for nuclear import or export, respectively. It has therefore been a major focus of recent research to detect cellular proteins that interact specifically with these domains in Rev and Rex, thereby potentially mediating Rev/Rex biological activity.

Nuclear import of proteins is generally mediated by NLSs, which are characterized by clusters of basic (frequently lysine) amino acid residues (for reviews, see DINGWALL and LASKEY 1991; BOULIKAS 1993). A prototypic NLS of this kind was originally identified in the SV40 large T-antigen. This type of NLS is recognized by a cytosolic heterodimeric protein complex. The importin-α subunit of this complex, also referred to as the "NLS receptor", binds directly to the NLS of the import substrate, while the importin-β subunit interacts with filaments that emanate from the cytoplasmic surface of the NPC. The import complex then moves to the central channel and is translocated through the NPC and into the nucleus (for general reviews on nuclear import, see BOULIKAS 1993; DINGWALL and LASKEY 1991; CORBETT and SILVER 1997; GÖRLICH and KUTAY 1999). In the case of HIV-1 Rev, however, it was initially demonstrated by in vitro binding studies that the importin-α subunit does not bind to the Rev NLS. In fact, it is the importin-β subunit which directly interacts with the arginine-rich Rev NLS (HENDERSON and PERCIPALLE 1997). The observation that importin-β can mediate nuclear import of the Rev as well as the Rex NLS was subsequently demonstrated in functional import assays using digitonin-permeabilized cell systems (TRUANT and CULLEN 1999; PALMERI and MALIM 1999). In the nucleus, high levels of RanGTP result in the dissociation of the complex of importin-β with Rev or Rex. As a consequence, both NLSs then become accessible for interaction with their respective viral RNA target sequences, eventually leading to nuclear export (Fig. 3).

An important, although unresolved question is, whether or not Rev/Rex has to move to a specific nuclear subcompartment in order to join viral RNAs. Although Rev and Rex are permanently shuttling, it is intriguing that both proteins predominantly accumulate in the nucleoli (COCHRANE et al. 1990b; MALIM et al. 1989b; KUBOTA et al. 1992; DUNDR et al. 1995; HOFER et al. 1991; NOSAKA et al. 1989; SIOMI et al. 1988). It has therefore been speculated that nucleolar localization might be important for biological Rev/Rex activity. In vitro binding

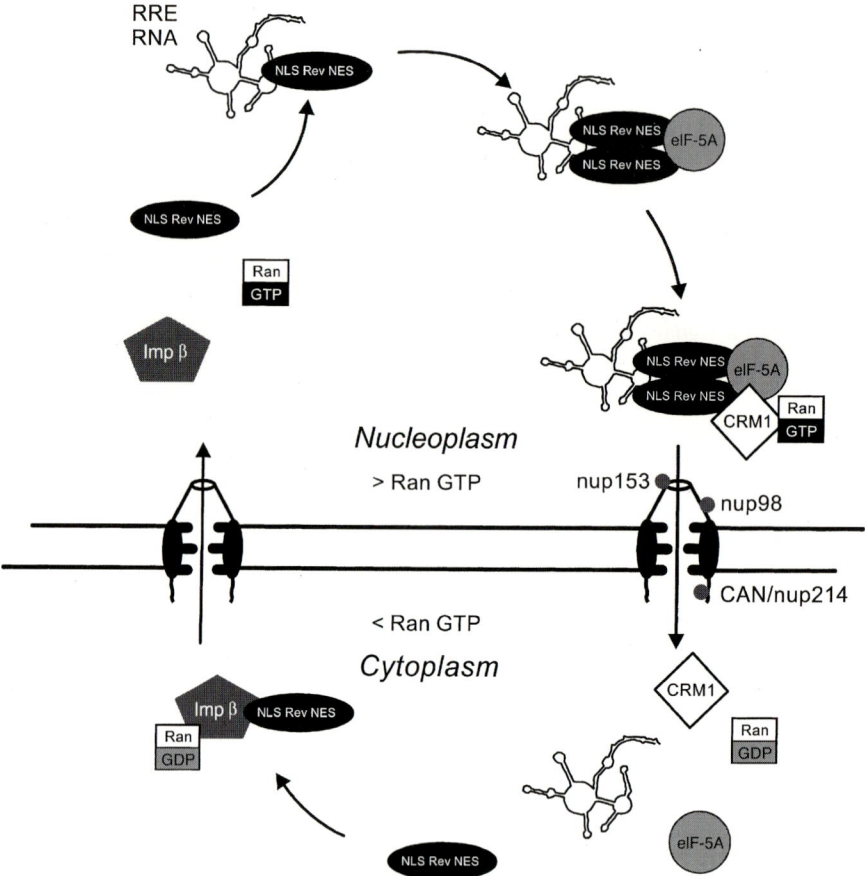

Fig. 3. Model of HIV-1 Rev action. In the nucleus, Rev binds to the RRE RNA target sequence present on all unspliced and single-spliced viral mRNAs. Rev binds directly to the RRE as a monomer via its RNA-binding domain/NLS. Subsequently, protein–protein interactions result in the binding of multimeric Rev complexes to the RRE. Formation of a stable Rev dimer appears to be required for the interaction of eIF-5A with the Rev activation domain/NES. The resulting RNP then binds via the Rev NES to the general export receptor CRM1/exportin 1 and moves to the nucleoplasmic face of the nuclear pore complex (NPC). Importantly, stable complex formation of the Rev NES with CRM1/exportin 1 depends on the presence of RanGTP, which is present at high levels in the nucleus. Translocation of the Rev/RRE-containing export substrate through the NPC is mediated by CRM1/exportin 1, which involves repeated docking and undocking events to structural components of the NPC (nucleoporins). The nucleoporins nup153, nup98 and CAN/nup214 have been shown to participate in Rev NES-mediated nuclear export. In the cytoplasm, conversion of RanGTP to RanGDP causes the dissociation of the nuclear export complex. Monomeric Rev is reimported into the nucleus by direct binding of the Rev NLS to importin-β and RanGDP. Following translocation into the nucleus, the exchange of GDP for GTP results in the disassembly of the Rev-containing import complex. Rev re-enters another nuclear export cycle by binding to the RRE RNA

studies have demonstrated that Rev forms a complex via its NLS/RBD with the nucleolar shuttle protein B23 and that this complex can be dissociated upon addition of RRE RNA (FANKHAUSER et al. 1991). However, no functional evidence exists that the complex formation observed in vitro is indeed required for Rev *trans*-activation activity. In fact, a recent study demonstrated that Rev binds to cellular 5S rRNA and its homologous, primary, RRE RNA-binding site in a similar fashion (LAM et al. 1998). Thus, the obvious affinity of the viral Rev *trans*-activator protein for 5S rRNA might account for its pronounced nucleolar accumulation. The notion that nucleolar localization is unnecessary for Rev/Rex function was originally suggested in studies by MCDONALD et al. (1992) and VENKATESAN et al. (1992), in which Rev was fused in-frame to a heterologous RNA-binding domain derived from the MS2 phage coat protein. This experimental design allowed the monitoring of *trans*-activation rates using a chimeric reporter construct that encoded the MS2 operator RNA sequence. Interestingly, mutations in the Rev NLS that abrogated nucleolar accumulation of these fusion proteins did not negatively affect their biological activity (MCDONALD et al. 1992; VENKATESAN et al. 1992). While these data suggested that nucleolar localization is not required for Rev function, it is sometimes argued that the data leading to these results were raised by using a heterologous RNA reporter which may accumulate in a different nuclear subcompartment than the authentic viral RNA. In situ RNA hybridization experiments have, however, shown that Rev promotes the nuclear export of RRE-containing RNAs that are dispersed throughout the nucleoplasm, exluding the nucleoli (ZHANG et al. 1996).

Another cellular factor that binds to the Rev NLS was originally identified by yeast two-hybrid screening (LUO et al. 1994; TANGE et al. 1996). Since this protein, p32, is associated with the alternative splicing factor SF2/ASF, it may function as a link between Rev and the cellular splicing machinery. Rev has been shown to mediate nuclear export not only of spliceable but, importantly, also of nonspliceable RRE-containing RNAs (FISCHER et al. 1994, 1999). Therefore Rev acts primarily at the level of nuclear export of RRE-containing RNAs and any potential effects on the splicing of viral mRNAs appear to be indirect.

Various experimental approaches have identified a series of cellular proteins that either directly or indirectly interact with the activation domain/NES of Rev and Rex. To date, the most prominent factor is CRM1/exportin 1, a protein that was originally identified in yeast, where its mutation affects chromosome morphology (chromosome region maintenance) (ADACHI and YANAGIDA 1989). Since then, multiple studies have provided evidence that CRM1/exportin 1 is a bona fide nuclear export receptor that mediates the nucleocytoplasmic translocation of the Rev/Rex cargo through the NPC (for review, see OHNO et al. 1998; GÖRLICH and KUTAY 1999; NAKIELNY and DREYFUSS 1999; STUTZ and ROSBASH 1998; WEIS 1998). CRM1/exportin 1 has a significant homology to importin-β (GÖRLICH et al. 1997), localizes in the nucleus as well as at the nucleoplasmic and cytoplasmic surface of NPCs (FORNEROD et al. 1997a), and associates particularly with the NPC component CAN/nup 214 (FORNEROD et al. 1996). Leptomycin B, which is a specific low-molecular-mass inhibitor of CRM1/exportin 1 (KUDO et al. 1998, 1999) has subsequently been used to demonstrate that CRM1/exportin 1 not only mediates nucleocytoplasmic translocation of Rev and Rex (WOLFF et al. 1997; FORNEROD et al. 1997b; FUKUDA et al. 1997; ELFGANG et al. 1999), but also the nuclear export of any other leucine-rich NES-containing proteins that have been reported to date. Because this permanently growing list of viral and cellular export substrates contains proteins with completely different biological activities (e.g., RNA transport factors, cellular transcription factors and their inhibitors, protein kinases and inhibitors thereof), it is fair to assume that, in addition to CRM1/exportin 1, other factors are required to achieve specific nuclear export regulation of these different export cargoes. In vitro experiments have demonstrated that CRM1/exportin 1 only interacts with the Rev NES in the presence of RanGTP (FORNEROD et al. 1997b; STADE et al. 1997) (Fig. 3). The requirement for RanGTP in order to form a stable Rev NES-CRM1/exportin 1 complex appears also to determine the direction of nuclear export (FISCHER et al. 1999), since RanGTP levels are high in the nucleus and low in the cytoplasm (for reviews of Ran function, see MOORE and BLOBEL 1994; MELCHIOR and GERACE 1998; MOORE 1998). Quantitative studies using RanGTP and NESs of different origin have revealed that significant differences exist with respect to NES-CRM1/exportin 1 binding affinities (ASKJAER et al. 1999). For example, a leucine-rich NES derived from protein kinase inhibitor (PKI) displayed strong affinity for CRM1/exportin 1, while, in sharp contrast, the Rev NES was characterized by an extremely low affinity for CRM1/exportin 1, at a level that was almost indistinguishable from that of an export-deficient mutant NES (ASKJAER et al. 1999). The combined data demonstrated that CRM1/exportin 1, together with RanGTP, is essential for the translocation of Rev/Rex NES-containing cargo through the NPC, but it also indicated that additional bridging factors or adapters are potentially required for efficient NES-CRM1/exportin 1 interaction.

A human protein, termed hRIP/Rab, and a similar yeast factor, Rip1p, were identified as other possible targets of the Rev/Rex NES in yeast two-hybrid assays (FRITZ et al. 1995; BOGERD et al. 1995; STUTZ et al. 1995). hRIP/Rab and Rip1p are associated with the NPC and are structurally characterized by repeated FG amino acid motifs which are typical of nucleoporins. In vitro binding studies using recombinant proteins, however, failed to demonstrate a direct interaction between the Rev NES and hRIP/Rab/Rip1p (STUTZ et al. 1996). It has since been shown that the initially observed "binding" of the Rev NES to Rip1p in yeast two-hybrid experiments is actually mediated by Crm1p (NEVILLE et al. 1997), the yeast homologue of CRM1/exportin 1 (STADE et al. 1997). Thus, during transit through the NPC, CRM1/exportin 1 appears to mediate the indirect contact of the NES-containing export cargo with multiple NPC components via its strong affinity for nucleoporin FG-repeat domains (STUTZ et al. 1996; FRITZ and GREEN 1996; FORNEROD et al. 1997a). However, when the *RIP1* gene in yeast was knocked out, it did not abrogate Rev function (STUTZ et al. 1995, 1997), and microinjection of anti-hRIP/Rab antibodies into the nucleus of somatic cells did not prevent export of nuclear injected GST-Rev protein (SCHATZ et al. 1998). These functional data indicate that hRIP/Rab is not an essential cellular cofactor for the Rev/Rex class of retroviral *trans*-activator proteins. Notably, previous studies in yeast have demonstrated that Rip1p is a factor that participates in the nuclear export of heat shock mRNAs (SAAVEDRA et al. 1997; STUTZ et al. 1997).

Another Rev/Rex cofactor candidate was identified by biochemical cross-linking experiments. The use of Rev activation domain peptides and reversible cross-linker molecules identified eukaryotic initiation factor 5A (eIF-5A) as a nuclear protein that interacts with the Rev activation domain (RUHL et al. 1993). eIF-5A is unique, because it is the only cellular protein known to date to contain the unusual amino acid hypusine (PARK et al. 1993). The hypusine modification is a spermidine-dependent posttranslational modification that is essential for eIF-5A function. Hypusine formation is catalyzed by the action of the two enzymes, deoxyhypusine synthase and deoxyhypusine hydroxylase (PARK et al. 1993). The finding that eIF-5A also localizes to the nucleus was rather surprising, since eIF-5A was historically considered to be an exclusively cytoplasmic factor involved in the initiation of protein synthesis (KEMPER et al. 1976; BENNE et al. 1978). Subsequent functional studies have, however, demonstrated that the designation of eIF-5A as an "initiation factor" is misleading, since intracellular depletion of eIF-5A in yeast has no negative effect on protein synthesis (KANG and HERSHEY 1994; KANG et al. 1993). A detailed analysis of eIF-5A localization in somatic cells and *Xenopus* oocytes by indirect immunofluorescence and immunogold electron microscopy has demonstrated that eIF-5A localizes in both the cytoplasm and nucleus and accumulates to some extent at the nucleoplasmic site of NPCs (ROSORIUS et al. 1999). This association of eIF-5A with the NPC appears to take place at the nuclear-pore-attached filaments that emanate from the pore margin into the nucleoplasm. These filaments are thought to be the initial docking sites of export cargoes to the NPC. Thus, eIF-5A may play a role in the targeting of export substrates to the NPC for subsequent CRM1/exportin-1-mediated transport through the pore channel. This notion is supported by the fact that, upon microinjection into the cell nucleus, eIF-5A is transported to the cytoplasm in an energy-dependent manner and binds to CRM1/exportin 1 in vitro (ROSORIUS et al. 1999). The strongest functional evidence that eIF-5A is an essential cofactor of Rev and Rex, however, originated from the finding that constitutive expression of eIF-5A mutant proteins (e.g., M14) blocks Rev function and thereby HIV-1 replication in *trans* (BEVEC et al. 1996; JUNKER et al. 1996). Moreover, microinjection of eIF-5A M14 protein, but not eIF-5A wild-type protein, into the cell nucleus specifically inhibits Rev/Rex NES-mediated nuclear export (ELFGANG et al. 1999). More recent data have revealed that the Rev/Rex inhibitory phenotype of the eIF-5A M14 protein can be explained by its lack of binding to CRM1/exportin 1, thereby apparently failing to mediate efficient Rev NES-CRM1/exportin 1 interaction (M.-C. Dabauvalle and J. Hauber, unpublished observation). Thus, eIF-5A appears to act as Rev/Rex cofactor by increasing the affinity of the Rev/Rex NES for CRM1/exportin 1 and associated components (e.g., RanGTP) (Fig. 3). It is also interesting to point out that, unlike CRM1/exportin 1, eIF-5A does not appear to be a general export factor required for all leucine-rich NESs. For example, the eIF-5A M14 protein inhibits Rev/Rex NES-mediated export but not the nucleocytoplasmic translocation of PKI NES (ELFGANG et al. 1999). This again indicates that other factors, in addition to CRM1/exportin 1, contribute to the nuclear export of some export cargoes. Obviously, eIF-5A appears to serve as a specific adapter or bridging factor in Rev/Rex-mediated nuclear export. This notion is also in agreement with the observation, mentioned previously, that CRM1/exportin 1 has an extremely poor binding affinity for the Rev NES (ASKJAER et al. 1999) and the recent finding that, in somatic cells, leucine-rich NESs derived from different export cargoes display completely different export kinetics. These export kinetics range from a few minutes up to several hours until translocation from the nucleus to the cytoplasm is complete (HENDERSON and ELEFTHERIOU 2000).

It should, however, be noted that some discrepancies exist with respect to Rev NES-eIF-5A binding. As mentioned above, eIF-5A was originally identified as a potential Rev cofactor by using structurally flexible Rev NES peptide mimics (RUHL et al. 1993). In contrast, when full-length Rev protein was used in in vitro binding experiments, an interaction with eIF-5A was only observed when Rev was bound to the RRE RNA at concentrations that allowed the formation of dimeric Rev-specific complexes (BEVEC et al. 1996). In contrast, no interaction between Rev and eIF-5A was detected in another study in which Rev/RRE ratios were used that favored the formation of monomeric or multimeric Rev-specific complexes on the RRE (HENDERSON and PERCIPALLE 1997). Clearly more in vitro binding experiments in which Rev is carefully titrated are required to resolve this open question or discrepancy. Nevertheless, these data indicate that the formation of a Rev homodimer on the RRE RNA might affect the structure of the Rev carboxy-terminus, thereby allowing or facilitating interaction of the Rev NES with cellular cofactors. This appears to be an attractive hypothesis, since it would integrate the requirements of RNA recognition and multimer formation with their subsequent activities in nuclear export in a more complete model of Rev/Rex activity.

It should also be noted that, to date, the number of Rev molecules per RRE RNA that are sufficient to form an export-competent complex has not been convincingly determined. As indicated before, this appears to be due to the extremely sticky nature of the Rev protein, which makes it very difficult to distinguish between functionally relevant and unspecific Rev complexes using in vitro binding studies. Interestingly, a study in which Rev multimerization in vitro was investigated in combination with functional cell-based assays concluded that two monomeric Rev proteins bound to the RRE are sufficient for biological activity (DALY et al. 1993b). The notion that RNA binding and multimerization of Rev may trigger a conformational change in the activation domain and is therefore a prerequisite of the interaction of Rev with cellular nuclear export factors, makes perfect sense with respect to viral RNA export. The fact that, in somatic cells and *Xenopus* oocytes, microinjected GST-Rev proteins are exported from the nucleus to the cytoplasm in the absence of RRE RNA does not actually conflict with this hypothesis, since highly abundant and structured 5 S rRNA can, as discussed before, act as a RNA target for Rev on which Rev multimer formation can occur (LAM et al. 1998; ROSORIUS et al. 2000). As shown in multiple studies, the requirement of RNA binding and multimerization can also be overcome by using flexible Rev/Rex NES peptides or GST-Rev NES fusion proteins in which, for example, a structurally flexible glycine linker element separates the GST from the NES moiety (ELFGANG et al. 1999).

A central step in the transport of Rev/Rex-containing ribonucleoprotein particles is their translocation through the NPC, a process that generally involves multiple docking and undocking events of the transport complexes to structural NPC components. As discussed before, it has been shown that FG-repeat domains of various nuleoporins interact indirectly with the Rev NES via CRM1/exportin 1 (STUTZ et al. 1996; FRITZ and GREEN 1996; NEVILLE et al. 1997). It was therefore not surprising to find that the nucleoporin CAN/nup214, which binds CRM1/exportin 1 via its carboxy-terminal FG-repeat region (FORNEROD et al. 1996, 1997a), participates in Rev/Rex NES-mediated nuclear export (ZOLOTUKHIN and FELBER 1999; BOGERD et al. 1998). CAN/nup214 accumulates primarily at the cytoplasmic surface of NPC (KRAEMER et al. 1994). However, when expressed alone, the CAN/nup214 carboxy-terminus localizes to the nucleoplasm and, importantly, causes the disappearance of CRM1/exportin 1 and eIF-5A from the NPC (FORNEROD et al. 1996; ROSORIUS et al. 1999). In turn, this also results in effective inhibition of nuclear export of Rev and Rex or Rev/Rex-dependent gene expression (ZOLOTUKHIN and FELBER 1999; BOGERD et al. 1998). In addition to CAN/nup214, two other nucleoporins, nup98 and nup153, which both localize to the nucleoplasmic periphery of NPC and are known to participate in nuclear export of multiple RNAs (snRNA, mRNA, 5S rRNA) (POWERS et al. 1997; BASTOS et al. 1996), have been shown to be involved in Rev NES-mediated export (ZOLOTUKHIN and FELBER 1999; ULLMAN et al. 1999) (Fig. 3). As in the case of CAN/nup214, the expression of the isolated FG-repeat domain of nup98 has been shown to inhibit Rev-mediated HIV-1 expression in human cells (ZOLOTUKHIN and FELBER 1999), and microinjection of antibodies directed against nup153 have been shown to block Rev export and Rev-mediated RRE RNA export in *Xenopus* oocytes (ULLMAN et al. 1999).

5 Perspectives

The Rev and Rex regulatory proteins provide an important mechanistic link by which human retroviruses exploit a specific, cellular RNA nuclear export pathway. Investigation of Rev/Rex function has already provided numerous insights into the mechanism of intracellular trafficking and identified various cellular proteins that participate in these processes. Nevertheless, many aspects of Rev/Rex function are still unknown. As mentioned before, it is still unclear how many Rev/Rex monomers bound to a single RRE or RxRE RNA molecule are actually sufficient to form a nuclear export competent RNP. The answer to this question would allow the requirement of Rev/Rex multimerization for functional activity to be integrated into a comprehensive model of Rev/Rex-mediated nuclear export. Another unresolved aspect of Rev function particularly is the recent finding that Rev activity is significantly diminished in human astrocytes (NEUMANN et al. 1995; LUDWIG et al. 1999), which are target cells for HIV-1 in the central nervous system. It remains to be seen precisely which functional aspect of Rev activity (e.g., nuclear import, nuclear export or nuclear retention) is impaired in these specialized cells.

Relatively little is known about which specific subnuclear compartments Rev and Rex must localize to in order to meet their RNA export cargo (e.g., compartments where RNA processing takes place) and to exercise their *trans*-activation activity. Our limited knowledge on this topic is reflected by sometimes conflicting reports on the Rev-dependent intranuclear localization of RRE-containing mRNAs and whether or not this distinct localization is associated with splicing component-35 (SC-35) containing nuclear granules (ZHANG et al. 1996; FAVARO et al. 1998; LUZNIK et al. 1995; SEGUIN et al. 1998; BOE et al. 1998; BERTHOLD and MALDARELLI 1996).

Since Rev-responsive viral mRNAs are clearly retained in the nucleus in the absence of Rev, it would also be interesting to examine the structural requirements that are necessary to trap these messages in the nucleus in more detail. The identification of cellular factors that bind to these viral RNAs may provide novel ways to experimentally approach these largely unresolved questions. In that respect, it is important to note that, for example, various heterogeneous nuclear ribonucleoproteins (hnRNPs) have been reported to interact with HIV-1 CRS/INS elements present on Rev-responsive viral mRNAs (BLACK et al. 1996; OLSEN et al. 1992; NAJERA et al. 1999). However, the exact functional contribution of these interactions to distinct subnuclear localization of the bound viral RNAs or to Rev-mediated nuclear export remains to be established. Another interesting cellular factor that has been reported to bind directly to the RRE is Sam68 (Src-associated protein in mitosis), which can to some extent substitute for or synergize with Rev in RRE-mediated gene expression (REDDY et al. 1999). Again, the contribution of Sam68 to Rev function in the context of the viral life cycle is relatively obscure since Rev-deficient mutant viruses are replication-incompetent.

As discussed before, the nuclear export pathway that is exploited by Rev and Rex appears to overlap with a specific RNA export pathway, since inhibition of this pathway not only blocks Rev export but also nucleocytoplasmic translocation of 5S rRNA and U snRNA (FISCHER et al. 1995). It will be of particular interest to dissect this pathway further in order to identify potentially novel soluble factors that are associated with Rev/Rex-containing RNP particles and which may be involved in the fine-regulation of this specific nucleocytoplasmic RNA transport system. The detailed investigation of the cellular cofactors involved in Rev/Rex export leads to another interesting, but also largely unresolved question: What is the exact function of the unique hypusine modification in eIF-5A? In this context it is interesting to note that eIF-5A mutant proteins (e.g., M14) are clearly inhibitory with respect to Rev/Rex function and HIV-1 replication when they are constitutively expressed in human cells (BEVEC et al. 1996; JUNKER et al. 1996). In sharp contrast, however, they have only little Rev/Rex-inhibitory effect when overexpressed in transient assays (J. Hauber, unpublished observation). This appears to be due to the fact that the hypusine modification enzymes (deoxyhypusine synthase and deoxyhypusine hydroxylase) are present in limiting amounts in the cell and are overwhelmed by large amounts of unmodified eIF-5A progenitor protein. Therefore, overexpression of eIF-5A results predominantly in the production of inactive eIF-5A. Moreover, injection of unmodified eIF-5A into the nucleus of somatic cells results in its export to the cytoplasm (ROSORIUS et al. 1999), and injected (unmodified) mutant proteins (e.g., eIF-5A M14) abrogate nuclear export of Rev and Rex (ELFGANG et al. 1999). These combined data suggest that the hypusine modification in eIF-5A is required for a so-far-unknown, although essential, eIF-5A activity that takes place prior to nuclear export. The finding that inhibitors of dexoyhypusine hydroxylase interfere with HIV-1 replication also reinforces the notion that this unique modification is required for the function of eIF-5A as a Rev/Rex cofactor (ANDRUS et al. 1998).

In summary, it is expected that the ongoing investigation into the various aspects of Rev/Rex function will not only continue to accelerate our knowledge on the regulation of nuclear export, but will hopefully also provide novel opportunities for pharmacological intervention in the HIV and HTLV life cycles, based on interference with Rev/Rex-mediated nuclear export.

Acknowledgements. I thank Prof. Marie-Christine Dabauvalle and Dr. Dorian Bevec for their collaboration and many helpful discussions on Rev and eIF-5A function throughout the years, Alexander Prechtel for help with the figures, and Dr. Sarah L. Thomas for critical comments on the manuscript.

References

Adachi Y, Yanagida M (1989) Higher order chromosome structure is affected by cold-sensitive mutations in a *Schizosaccharomyces pombe* gene crm1 + which encodes a 115-kD protein preferentially localized in the nucleus and its periphery. J Cell Biol 108:1195–1207

Ahmed YF, Gilmartin GM, Hanly SM, Nevins JR, Greene WC (1991) The HTLV-1 Rex response element mediates a novel form of mRNA polyadenylation. Cell 64:727–737

Ahmed YF, Hanly SM, Malim MH, Cullen BR, Greene WC (1990) Structure-function analyses of the HTLV-1 Rex and HIV-1 Rev RNA response elements: insights into the mechanism of Rex and Rev action. Genes Dev 4:1014–1022

Andrus L, Szabo P, Grady RW, Hanauske AR, Huima Byron T, Slowinska B, Zagulska S, Hanauske Abel HM (1998) Antiretroviral effects of deoxyhypusyl hydroxylase inhibitors: a hypusine-dependent host cell mechanism for replication of human immunodeficiency virus type 1 (HIV-1). Biochem Pharmacol 55:1807–1818

Askjaer P, Bachi A, Wilm M, Bischoff FR, Weeks DL, Ogniewski V, Ohno M, Niehrs C, Kjems J, Mattaj IW, Fornerod M (1999) RanGTP-regulated interactions of CRM1 with nucleoporins and a shuttling DEAD-box helicase. Mol Cell Biol 19:6276–6285

Askjaer P, Kjems J (1998) Mapping of multiple RNA binding sites of human T-cell lymphotropic virus type I Rex protein within 5′- and 3′-Rex response elements. J Biol Chem 273:11463–11471

Auer M, Gremlich H-U, Seifert J-M, Daly TJ, Parslow TG, Casari G, Gstach H (1994) Helix-loop-helix motif in HIV-1 Rev. Biochemistry 33:2988–2996

Ballaun C, Farrington GK, Dobrovnik M, Rusche J, Hauber J, Böhnlein E (1991) Functional analysis of human T-cell leukemia virus type I rex-response element: Direct RNA binding of Rex protein correlates with in vivo activity. J Virol 65:4408–4413

Baskerville S, Zapp M, Ellington AD (1995) High-resolution mapping of the human T-cell leukemia virus type 1 Rex-binding element by in vitro selection. J Virol 69:7559–7569

Bastos R, Lin A, Enarson M, Burke B (1996) Targeting and function in mRNA export of nuclear pore complex protein Nup153. J Cell Biol 134:1141–1156

Benne R, Brown-Luedi ML, Hershey JW (1978) Purification and characterization of protein synthesis initiation factors eIF-1, eIF-4 C, eIF-4D, and eIF-5 from rabbit reticulocytes. J Biol Chem 253:3070–3077

Berthold E, Maldarelli F (1996) Cis-acting elements in human immunodeficiency virus type 1 RNAs direct viral transcripts to distinct intranuclear locations. J Virol 70:4667–4682

Bevec D, Jaksche H, Oft M, Wöhl T, Himmelspach M, Pacher A, Schebesta M, Koettnitz K, Dobrovnik M, Csonga R, Lottspeich F, Hauber J (1996) Inhibition of HIV-1 replication in lymphocytes by mutants of the Rev cofactor eIF-5A. Science 271:1858–1860

Black AC, Luo J, Chun S, Bakker A, Fraser JK, Rosenblatt JD (1996) Specific binding of polypyrimidine tract binding protein and hnRNP A1 to HIV-1 CRS elements. Virus Genes 12:275–285

Boe SO, Bjorndal B, Rosok B, Szilvay AM, Kalland KH (1998) Subcellular localization of human immunodeficiency virus type 1 RNAs, Rev, and the splicing factor SC-35. Virology 244:473–482

Bogerd H, Greene WC (1993) Dominant negative mutants of human T-cell leukemia virus type i Rex and human immunodeficiency virus type 1 Rev fail to multimerize in vivo. J Virol 67:2496–2502

Bogerd HP, Echarri A, Ross TM, Cullen BR (1998) Inhibition of human immunodeficiency virus Rev and human T-cell leukemia virus Rex function, but not Mason-Pfizer monkey virus constitutive transport element activity, by a mutant human nucleoporin targeted to Crm1. J Virol 72:8627–8635

Bogerd HP, Fridell RA, Benson RE, Hua J, Cullen BR (1996) Protein sequence requirements for function of the human T-cell leukemia virus type 1 Rex nuclear export signal delineated by a novel in vivo randomization-selection assay. Mol Cell Biol 16:4207–4214

Bogerd HP, Fridell RA, Madore S, Cullen BR (1995) Identification of a novel cellular cofactor for the Rev/Rex class of retroviral regulatory proteins. Cell 82:485–494

Bogerd HP, Huckaby GL, Ahmed YS, Hanly SM, Greene WC (1991) The type I human T-cell leukemia virus (HTLV-1) Rex trans-activator binds directly to the HTLV-1 Rex and the type 1 human immunodeficiency virus Rev RNA response elements. Proc Natl Acad Sci USA 88:5704–5708

Boulikas T (1993) Nuclear localization signals (NLS). Crit Rev Eukaryot Gene Expr 3:193–227

Böhnlein E, Berger J, Hauber J (1991a) Functional mapping of the human immunodeficiency virus type 1 Rev RNA binding domain: new insights into the domain structure of Rev and Rex. J Virol 65:7051–7055

Böhnlein S, Pirker FP, Hofer L, Zimmermann K, Bachmayer H, Böhnlein E, Hauber J (1991b) Trans-dominant repressors for human T-cell leukemia virus type I Rex and human immunodeficiency virus type 1 Rev function. J Virol 65:81–88

Brice PC, Kelley AC, Butler PJ (1999) Sensitive in vitro analysis of HIV-1 Rev multimerization. Nucleic Acids Res 27:2080–2085

Chang DD, Sharp PA (1989) Regulation by HIV Rev depends upon recognition of splice sites. Cell 59:789–795

Cochrane AW, Chen C-H, Rosen CA (1990a) Specific interaction of the human immunodeficiency virus Rev protein with a structured region in the env mRNA. Proc Natl Acad Sci USA 87:1198–1202

Cochrane AW, Jones KS, Beidas S, Dillon PJ, Skalka AM, Rosen CA (1991) Identification and characterization of intragenic sequences which repress HIV structural gene expression. J Virol 65:5305–5313

Cochrane AW, Perkins A, Rosen CA (1990b) Identification of sequences important in the nucleolar localization of human immunodeficiency virus Rev: relevance of nucleolar localization to function. J Virol 64:881–885

Cole JL, Gehman JD, Shafer JA, Kuo LC (1993) Solution oligomerization of the Rev protein of HIV-1: Implications for function. Biochemistry 32:11769–11775

Cook KS, Fisk GJ, Hauber J, Usman N, Daly TJ, Rusche JR (1991) Characterization of HIV-1 REV protein: binding stoichiometry and minimal RNA substrate. Nucleic Acids Res 19:1577–1583

Corbett AH, Silver PA (1997) Nucleocytoplasmic transport of macromolecules. Microbiol Mol Biol Rev 61:193–211

Cullen BR, Hauber J, Campbell K, Sodroski JG, Haseltine WA, Rosen CA (1988) Subcellular localization of the human immunodeficiency virus trans-acting art gene product. J Virol 62:2498–2501

Daefler S, Klotman ME, Wong-Staal F (1990) Trans-activating rev protein of the human immunodeficiency virus 1 interacts directly and specifically with its target RNA. Proc Natl Acad Sci USA 87:4571–4575

Daly TJ, Cook KS, Gray GS, Maione TE, Rusche JR (1989) Specific binding of HIV-1 recombinant Rev protein to the Rev-responsive element in vitro. Nature 342:816–819

Daly TJ, Doten RC, Rennert P, Auer M, Jaksche H, Donner A, Fisk G, Rusche JR (1993a) Biochemical characterization of binding of multiple HIV-1 Rev monomeric proteins to the Rev responsive element. Biochemistry 32:10497–10505

Daly TJ, Rennert P, Lynch P, Barry JK, Dundas M, Rusche JR, Doten RC, Auer M, Farrington GK (1993b) Perturbation of the carboxy terminus of HIV-1 Rev affects multimerization on the Rev responsive element. Biochemistry 32:8945–8954

Dingwall C, Laskey RA (1991) Nuclear targeting sequences–a consensus? Trends Biochem Sci 16:478–481

Dundr M, Leno GH, Hammarskjöld M-L, Rekosh D, Helga-Maria C, Olson MOJ (1995) The roles of nucleolar structure and function in the subcellular location of the HIV-1 Rev protein. J Cell Sci 108:2811–2823

Elfgang C, Rosorius O, Hofer L, Jaksche H, Hauber J, Bevec D (1999) Evidence for specific nucleo-cytoplasmic transport pathways used by leucine-rich nuclear export signals. Proc Natl Acad Sci USA 96:6229–6234

Fankhauser C, Izaurralde E, Adachi Y, Wingfield P, Laemmli UK (1991) Specific complex of human immunodeficiency virus type 1 rev and nucleolar B23 proteins: dissociation by the Rev response element. Mol Cell Biol 11:2567–2575

Favaro JP, Borg KT, Arrigo SJ, Schmidt MG (1998) Effect of Rev on the intranuclear localization of HIV-1 unspliced RNA. Virology 249:286–296

Fischer U, Huber J, Boelens WC, Mattaj IW, Lührmann R (1995) The HIV-1 Rev activation domain is a nuclear export signal that accesses an export pathway used by specific cellular RNAs. Cell 82:475–483

Fischer U, Meyer S, Teufel M, Heckel C, Lührmann R, Rautmann G (1994) Evidence that HIV-1 Rev directly promotes the nuclear export of unspliced RNA. EMBO J 13:4105–4112

Fischer U, Pollard VW, Luhrmann R, Teufel M, Michael MW, Dreyfuss G, Malim MH (1999) Rev-mediated nuclear export of RNA is dominant over nuclear retention and is coupled to the Ran-GTPase cycle. Nucleic Acids Res 27:4128–4134

Fornerod M, Boer J, van Baal S, Morreau H, Grosveld G (1996) Interaction of cellular proteins with the leukemia specific fusion proteins DEK-CAN and SET-CAN and their normal counterpart, the nucleoporin CAN. Oncogene 13:1801–1808

Fornerod M, Ohno M, Yoshida M, Mattaj IW (1997b) CRM1 is an export receptor for leucine-rich nuclear export signals. Cell 90:1051–1060

Fornerod M, van Deursen J, van Baal S, Reynolds A, Davis D, Murti KG, Fransen J, Grosveld G (1997a) The human homologue of yeast CRM1 is in a dynamic subcomplex with CAN/Nup214 and a novel nuclear pore component Nup88. EMBO J 16:807–816

Fritz CC, Green MR (1996) HIV Rev uses a conserved cellular protein export pathway for the nucleo-cytoplasmic transport of viral RNAs. Curr Biol 6:848–854

Fritz CC, Zapp ML, Green MR (1995) A human nucleoporin-like protein that specifically interacts with HIV Rev. Nature 376:530–533

Fukuda M, Asano S, Nakamura T, Adachi M, Yoshida M, Yanagida M, Nishida E (1997) CRM1 is responsible for intracellular transport mediated by the nuclear export signal. Nature 390:308–311

Görlich D, Dabrowski M, Bischoff FR, Kutay U, Bork P, Hartmann E, Prehn S, Izaurralde E (1997) A novel class of RanGTP binding proteins. J Cell Biol 138:65–80

Görlich D, Kutay U (1999) Transport between the cell nucleus and the cytoplasm. Annu Rev Cell Dev Biol 15:607–660

Gröne M, Hoffmann E, Berchtold S, Cullen BR, Grassmann R (1994) A single stem-loop structure within the HTLV-1 Rex response element is sufficient to mediate Rex activity in vivo. Virology 204:144–152

Hadzopoulou-Cladaras M, Felber BK, Cladaras C, Athanassopoulos A, Tse A, Pavlakis GN (1989) The rev (trs/art) protein of human immunodeficiency virus type 1 affects viral mRNA and protein expression via a cis-acting sequence in the env region. J Virol 63:1265–1274

Hammarskjöld M-L, Li H, Rekosh D, Prasad S (1994) Human immunodeficiency virus env expression becomes Rev-independent if the env region is not defined as an intron. J Virol 68:951–958

Hammerschmid M, Palmeri D, Ruhl M, Jaksche H, Weichselbraun I, Böhnlein E, Malim MH, Hauber J (1994) Scanning mutagenesis of the arginine-rich region of the human immunodeficiency virus type 1 Rev trans activator. J Virol 68:7329–7335

Hanly SM, Rimsky LT, Malim MH, Kim JH, Hauber J, Duc Dodon M, Le S-Y, Maizel JV, Cullen BR, Greene WC (1989) Comparative analysis of the HTLV-1 Rex and HIV-1 Rev trans-regulatory proteins and their RNA response elements. Genes Dev 3:1534–1544

Heaphy S, Dingwall C, Ernberg I, Gait MJ, Green SM, Karn J, Lowe AD, Singh M, Skinner MA (1990) HIV-1 regulator of virion expression (Rev) protein binds to an RNA stem-loop structure located within the Rev response element region. Cell 60:685–693

Heger P, Rosorius O, Koch C, Casari G, Grassmann R, Hauber J (1998) Multimer formation is not essential for nuclear export of human T-cell leukemia virus type 1 Rex trans-activator protein. J Virol 72:8659–8668

Henderson BR, Eleftheriou A (2000) A comparison of the activity, sequence specificity, and CRM1-dependence of different nuclear export signals. Exp Cell Res 256:213–224

Henderson BR, Percipalle P (1997) Interactions between HIV Rev and nuclear import and export factors: the Rev nuclear localisation signal mediates specific binding to human importin-beta. J Mol Biol 274:693–707

Hofer L, Weichselbraun I, Quick S, Farrington GK, Böhnlein E, Hauber J (1991) Mutational analysis of the human T-cell leukemia virus type I trans-acting rex gene product. J Virol 65:3379–3383

Hope TJ, Bond BL, McDonald D, Klein NP, Parslow TG (1991) Effector domains of human immunodeficiency virus type 1 Rev and human T-cell leukemia virus type I Rex are functionally interchangeable and share an essential peptide motif. J Virol 65:6001–6007

Hope TJ, Huang X, McDonald D, Parslow TG (1990) Steroid-receptor fusion of the human immunodeficiency virus type 1 Rev transactivator: mapping cryptic functions of the arginine-rich motif. Proc Natl Acad Sci USA 87:7787–7791

Iwai S, Pritchard C, Mann DA, Karn J, Gait MJ (1992) Recognition of the high affinity binding site in rev-response element RNA by the human immunodeficiency virus type-1 rev protein. Nucleic Acids Res 20:6465–6472

Izaurralde E, Adam S (1998) Transport of macromolecules between the nucleus and the cytoplasm. RNA 4:351–364

Junker U, Bevec D, Barske C, Kalfoglou C, Escaich S, Dobrovnik M, Hauber J, Böhnlein E (1996) Intracellular expression of cellular eIF-5A mutants inhibits HIV-1 replication in human T cells: a feasibility study. Hum Gene Ther 7:1861–1869

Kalland K-H, Szilvay AM, Brokstad KA, Sætrevik W, Haukenes G (1994) The human immunodeficiency virus type 1 Rev protein shuttles between the cytoplasm and nuclear compartments. Mol Cell Biol 14:7436–7444

Kang HA, Hershey JWB (1994) Effect of initiation factor eIF-5A depletion on protein synthesis and proliferation of Saccharomyces cerevisiae. J Biol Chem 269:3934–3940

Kang HA, Schwelberger HG, Hershey JWB (1993) Effect of initiation factor eIF-5A depletion on cell proliferation and protein synthesis. In: Brown AJP, Tuite MF, and McCarthy JEG (eds) Protein synthesis and targeting in yeast. NATO Series H: Cell Biol Springer Berlin Heidelberg New York pp 123–129

Kemper WM, Berry KW, Merrick WC (1976) Purification and properties of rabbit reticulocyte protein synthesis initiation factors M2Balpha and M2Bbeta. J Biol Chem 251:5551–5557

Kim FJ, Beeche AA, Hunter JJ, Chin DJ, Hope TJ (1996) Characterization of the nuclear export signal of human T-cell lymphotropic virus type 1 Rex reveals that nuclear export is mediated by position-variable hydrophobic interactions. Mol Cell Biol 16:5147–5155

Kjems J, Calnan BJ, Frankel AD, Sharp PA (1992) Specific binding of a basic peptide from HIV-1 Rev. EMBO J 11:1119–1129

Kraemer D, Wozniak RW, Blobel G, Radu A (1994) The human CAN protein, a putative oncogene product associated with myeloid leukemogenesis, is a nuclear pore complex protein that faces the cytoplasm. Proc Natl Acad Sci USA 91:1519–1523

Kubota S, Furuta R, Maki M, Hatanaka M (1992) Inhibition of human immunodeficiency virus type 1 Rev function by a Rev mutant which interferes with nuclear/nucleolar localization of Rev. J Virol 66:2510–2513

Kubota S, Nosaka T, Cullen BR, Maki M, Hatanaka M (1991) Effects of chimeric mutants of human immunodeficiency virus type 1 Rev and human T-cell leukemia virus type I Rex on nucleolar targeting signals. J Virol 65:2452–2456

Kudo N, Matsumori N, Taoka H, Fujiwara D, Schreiner EP, Wolff B, Yoshida M, Horinouchi S (1999) Leptomycin B inactivates CRM1/exportin 1 by covalent modification at a cysteine residue in the central conserved region. Proc Natl Acad Sci USA 96:9112–9117

Kudo N, Wolff B, Sekimoto T, Schreiner EP, Yoneda Y, Yanagida M, Horinouchi S, Yoshida M (1998) Leptomycin B inhibition of signal-mediated nuclear export by direct binding to CRM1. Exp Cell Res 242:540–547

Lam WC, Seifert JM, Amberger F, Graf C, Auer M, Millar DP (1998) Structural dynamics of HIV-1 Rev and its complexes with RRE and 5 S RNA. Biochemistry 37:1800–1809

Ludwig E, Silberstein FC, van Empel J, Erfle V, Neumann M, Brack-Werner R (1999) Diminished rev-mediated stimulation of human immunodeficiency virus type 1 protein synthesis is a hallmark of human astrocytes. J Virol 73:8279–8289

Luo Y, Yu H, Peterlin BM (1994) Cellular protein modulates effects of human immunodeficiency virus type 1 Rev. J Virol 68:3850–3856

Luznik L, Martone ME, Kraus G, Zhang Y, Xu Y, Ellisman MH, Wong-Staal F (1995) Localization of human immunodeficiency virus Rev in transfected and virus-infected cells. AIDS Res Hum Retroviruses 11:795–804

Madore SJ, Tiley LS, Malim MH, Cullen BR (1994) Sequence requirements for Rev multimerization in vivo. Virology 202:186–194

Maldarelli F, Martin MA, Strebel K (1991) Identification of post-transcriptionally active inhibitory sequences in HIV-1 RNA; novel level of gene regulation. J Virol 65:5732–5743

Malim MH, Böhnlein S, Hauber J, Cullen BR (1989b) Functional dissection of the HIV-1 Rev trans-activator–derivation of a trans-dominant repressor of Rev function. Cell 58:205–214

Malim MH, Cullen BR (1991) HIV-1 structural gene expression requires the binding of multiple Rev monomers to the viral RRE: implications for HIV-1 latency. Cell 65:241–248

Malim MH, Cullen BR (1993) Rev and the fate of pre-mRNA in the nucleus: implications for the regulation of RNA processing in eukaryotes. Mol Cell Biol 13:6180–6189

Malim MH, Hauber J, Le S-Y, Maizel JV, Cullen BR (1989a) The HIV-1 rev trans-activator acts through a structured target sequence to activate nuclear export of unspliced viral mRNA. Nature 338:254–257

Malim MH, McCarn DF, Tiley LS, Cullen BR (1991) Mutational definition of the human immunodeficiency virus type 1 Rev activation domain. J Virol 65:4248–4254

Malim MH, Tiley LS, McCarn DF, Rusche JR, Hauber J, Cullen BR (1990) HIV-1 structural gene expression requires binding of the Rev trans-activator to its RNA target sequence. Cell 60: 675–683

Mann DA, Mikaélian I, Zemmel RW, Green SM, Lowe AD, Kimura T, Singh M, Butler PJG, Gait MJ, Karn J (1994) A molecular rheostat. Co-operative Rev binding to Stem I of the Rev-response element modulates human immunodeficiency virus type-1 late gene expression. J Mol Biol 241:193–207

McDonald D, Hope TJ, Parslow TG (1992) Posttranscriptional regulation by the human immunodeficiency virus type 1 Rev and human T-cell leukemia virus type I Rex proteins through a heterologous RNA binding site. J Virol 66:7232–7238

Melchior F, Gerace L (1998) Two-way trafficking with Ran. Trends Cell Biol 8:175–179

Mermer B, Felber BK, Campbell M, Pavlakis GN (1990) Identification of trans-dominant HIV-1 rev protein mutants by direct transfer of bacterially produced proteins into human cells. Nucleic Acids Res 18:2037–2044

Meyer BE, Malim MH (1994) The HIV-1 Rev trans-activator shuttles between the nucleus and the cytoplasm. Genes Dev 8:1538–1547

Meyer BE, Meinkoth JL, Malim MH (1996) Nuclear transport of human immunodeficiency virus type 1, visna virus, and equine infectious anemia virus Rev proteins: Identification of a family of transferable nuclear export signals. J Virol 70:2350–2359

Moore MS (1998) Ran and nuclear transport. J Biol Chem 273:22857–22860

Moore MS, Blobel G (1994) A G protein involved in nucleocytoplasmic transport: the role of Ran. Trends Biochem Sci 19:211–216

Najera I, Krieg M, Karn J (1999) Synergistic stimulation of HIV-1 rev-dependent export of unspliced mRNA to the cytoplasm by hnRNP A1. J Mol Biol 285:1951–1964

Nakielny S, Dreyfuss G (1999) Transport of proteins and RNAs in and out of the nucleus. Cell 99:677–690

Nalin CM, Purcell RD, Antelman D, Mueller D, Tomchak L, Wegrzynski B, McCarney E, Toome V, Kramer R, Hsu MC (1990) Purification and characterization of recombinant Rev protein of human immunodeficiency virus type 1. Proc Natl Acad Sci USA 87:7593–7597

Neumann M, Felber BK, Kleinschmidt A, Froese B, Erfle V, Pavlakis GN, Brack-Werner R (1995) Restriction of human immunodeficiency virus type 1 production in a human astrocytoma cell line is associated with a cellular block in Rev function. J Virol 69:2159–2167

Neville M, Stutz F, Lee L, Davis LI, Rosbash M (1997) The importin-beta family member Crm1p bridges the interaction between Rev and the nuclear pore complex during nuclear export. Curr Biol 7:767–775

Nosaka T, Miyazaki Y, Takamatsu T, Sano K, Nakai M, Fujita S, Martin TE, Hatanaka M (1995) The post-transcriptional regulator Rex of the human T-cell leukemia virus type I is present as nucleolar speckles in infected cells. Exp Cell Res 219:122–129

Nosaka T, Siomi H, Adachi Y, Ishibashi M, Kubota S, Maki M, Hatanaka M (1989) Nucleolar targeting signal of human T-cell leukemia virus type I rex-encoded protein is essential for cytoplasmic accumulation of unspliced viral mRNA. Proc Natl Acad Sci USA 86:9798–9802

Ohno M, Fornerod M, Mattaj IW (1998) Nucleocytoplasmic transport: the last 200 nanometers. Cell 92:327–336

Olsen HS, Cochrane AW, Dillon PJ, Nalin CM, Rosen CA (1990b) Interaction of the human immunodeficiency virus type 1 Rev protein with a structured region in env mRNA is dependent on multimer formation mediated through a basic stretch of amino acids. Genes Dev 4:1357–1364

Olsen HS, Cochrane AW, Rosen C (1992) Interaction of cellular factors with intragenic *cis*-acting repressive sequences within the HIV genome. Virology 191:709–715

Olsen HS, Nelböck P, Cochrane AW, Rosen CA (1990a) Secondary structure is the major determinant for interaction of HIV rev protein with RNA. Science 247:845–848

Palmeri D, Malim MH (1996) The human T-cell leukemia virus type 1 posttranscriptional *trans*-activator Rex contains a nuclear export signal. J Virol 70:6442–6445

Palmeri D, Malim MH (1999) Importin beta can mediate the nuclear import of an arginine-rich nuclear localization signal in the absence of importin alpha. Mol Cell Biol 19:1218–1225

Park MH, Wolff EC, Folk JE (1993) Hypusine: its post-translational formation in eukaryotic initiation factor 5A and its potential role in cellular regulation. BioFactors 4:95–104

Pollard VW, Malim MH (1998) The HIV-1 Rev protein. Annu Rev Microbiol 52:491–532

Powell DM, Zhang MJ, Konings DAM, Wingfield PT, Stahl SJ, Dayton ET, Dayton AI (1995) Sequence specificity in the higher-order interaction of the Rev protein of HIV-1 with its target sequence, the RRE. J Acquir Immune Defic Syndr Hum Retrovirol 10:317–323

Powers MA, Forbes DJ, Dahlberg JE, Lund E (1997) The vertebrate GLFG nucleoporin, Nup98, is an essential component of multiple RNA export pathways. J Cell Biol 136:241–250

Reddy TR, Xu W, Mau JK, Goodwin CD, Suhasini M, Tang H, Frimpong K, Rose DW, Wong-Staal F (1999) Inhibition of HIV replication by dominant negative mutants of Sam68, a functional homolog of HIV-1 Rev [published erratum appears in Nat Med 1999 Jul; 5(7):849]. Nat Med 5:635–642

Rimsky L, Duc Dodon M, Dixon EP, Greene WC (1989) Trans-dominant inactivation of HTLV-1 and HIV-1 gene expression by mutation of the HTLV-1 Rex transactivator. Nature 341:453–456

Rimsky L, Hauber J, Dukovich M, Malim MH, Langlois A, Cullen BR, Greene WC (1988) Functional replacement of the HIV-1 rev protein by the HTLV-1 rex protein. Nature 335:738–740

Rosen CA, Terwilliger E, Dayton A, Sodroski JG, Haseltine WA (1988) Intragenic cis-acting art gene-responsive sequences of the human immunodeficiency virus. Proc Natl Acad Sci USA 85: 2071–2075

Rosorius O, Fries B, Stauber RH, Hirschmann N, Bevec D, Hauber J (2000) Human ribosomal protein L5 contains defined nuclear localization and export signals. J Biol Chem 275:12061–12068

Rosorius O, Reichart B, Krätzer F, Heger P, Dabauvalle MC, Hauber J (1999) Nuclear pore localization and nucleocytoplasmic transport of eIF-5A: evidence for direct interaction with the export receptor CRM1. J Cell Sci 112:2369–2380

Ruhl M, Himmelspach M, Bahr GM, Hammerschmid F, Jaksche H, Wolff B, Aschauer H, Farrington GK, Probst H, Bevec D, Hauber J (1993) Eukaryotic initiation factor 5A is a cellular target of the human immunodeficiency virus type 1 Rev activation domain mediating trans-activation. J Cell Biol 123:1309–1320

Saavedra CA, Hammell CM, Heath CV, Cole CN (1997) Yeast heat shock mRNAs are exported through a distinct pathway defined by Rip1p. Genes Dev 11:2845–2856

Schatz O, Oft M, Dascher C, Schebesta M, Rosorius O, Jaksche H, Dobrovnik M, Bevec D, Hauber J (1998) Interaction of the HIV-1 rev cofactor eukaryotic initiation factor 5A with ribosomal protein L5. Proc Natl Acad Sci USA 95:1607–1612

Schwartz S, Campbell M, Nasioulas G, Harrison J, Felber BK, Pavlakis GN (1992b) Mutational inactivation of an inhibitory sequence in human immunodeficiency virus type 1 results in Rev-independent gag expression. J Virol 66:7176–7182

Schwartz S, Felber BK, Pavlakis GN (1992a) Distinct RNA sequences in the gag region of HIV-1 decrease RNA stability and inhibit expression in the absence of the Rev protein. J Virol 66:150–159

Seguin B, Staffa A, Cochrane A (1998) Control of human immunodeficiency virus type 1 RNA metabolism: role of splice sites and intron sequences in unspliced viral RNA subcellular distribution. J Virol 72:9503–9513

Seiki M, Inoue J, Hidaka M, Yoshida M (1988) Two cis-acting elements responsible for posttranscriptional trans-regulation of gene expression of human T-cell leukemia virus type I. Proc Natl Acad Sci USA 85:7124–7128

Siomi H, Shida H, Nam SH, Nosaka T, Maki M, Hatanaka M (1988) Sequence requirements for nucleolar localization of human T cell leukemia virus type I pX protein, which regulates viral RNA processing. Cell 55:197–209

Stade K, Ford CS, Guthrie C, Weis K (1997) Exportin 1 (Crm1p) is an essential nuclear export factor. Cell 90:1041–1050

Staffa A, Cochrane A (1994) The tat/rev intron of human immunodeficiency virus type 1 is inefficiently spliced because of suboptimal signals in the 3′ splice site. J Virol 68:3071–3079

Stauber R, Gaitanaris GA, Pavlakis GN (1995) Analysis of trafficking of Rev and transdominant Rev proteins in living cells using green fluorescent protein fusions: Transdominant Rev blocks the export of Rev from the nucleus to the cytoplasm. Virology 213:439–449

Stutz F, Izaurralde E, Mattaj IW, Rosbash M (1996) A role for nucleoporin FG repeat domains in export of human immunodeficiency virus type 1 Rev protein and RNA from the nucleus. Mol Cell Biol 16:7144–7150

Stutz F, Kantor J, Zhang D, McCarthy T, Neville M, Rosbash M (1997) The yeast nucleoporin rip1p contributes to multiple export pathways with no essential role for its FG-repeat region. Genes Dev 11:2857–2868

Stutz F, Neville M, Rosbash M (1995) Identification of a novel nuclear pore-associated protein as a functional target of the HIV-1 Rev protein in yeast. Cell 82:495–506

Stutz F, Rosbash M (1998) Nuclear RNA export. Genes Dev 12:3303–3319

Tange TO, Jensen TH, Kjems J (1996) In vitro interaction between human immunodeficiency virus type 1 rev protein and splicing factor ASF/SF2-associated protein, p32. J Biol Chem 271:10066–10072

Thomas SL, Hauber J, Casari G (1997) Probing the structure of the HIV-1 Rev trans-activator protein by functional analysis. Protein Eng 10:103–107

Thomas SL, Oft M, Jaksche H, Casari G, Heger P, Dobrovnik M, Bevec D, Hauber J (1998) Functional analysis of the human immunodeficiency virus type 1 Rev protein oligomerization interface. J Virol 72:2935–2944

Tiley LS, Malim MH, Tewary HK, Stockley PG, Cullen BR (1992) Identification of a high-affinity RNA-binding site for the human immunodeficiency virus type 1 Rev protein. Proc Natl Acad Sci USA 89:758–762

Toyoshima H, Itoh M, Inoue J, Seiki M, Takaku F, Yoshida M (1990) Secondary structure of the human T-cell leukemia virus type 1 rex-responsive element is essential for rex regulation of RNA processing and transport of unspliced RNAs. J Virol 64:2825–2832

Truant R, Cullen BR (1999) The arginine-rich domains present in human immunodeficiency virus type 1 Tat and Rev function as direct importin beta-dependent nuclear localization signals. Mol Cell Biol 19:1210–1217

Ullman KS, Shah S, Powers MA, Forbes DJ (1999) The nucleoporin nup153 plays a critical role in multiple types of nuclear export. Mol Biol Cell 10:649–664

Unge T, Solomin L, Mellini M, Derse D, Felber BK, Pavlakis GN (1991) The Rex regulatory protein of human T-cell lymphotropic virus type I binds specifically to its target site within the viral RNA. Proc Natl Acad Sci USA 88:7145–7149

Venkatesan S, Gerstberger SM, Park H, Holland SM, Nam Y (1992) Human immunodeficiency virus type 1 Rev activation can be achieved without Rev-responsive element RNA if Rev is directed to the target as a Rev/MS2 fusion protein which tethers the MS2 operator RNA. J Virol 66:7469–7480

Venkatesh LK, Chinnadurai G (1990) Mutants in a conserved region near the carboxy-terminus of HIV-1 Rev identify functionally important residues and exhibit a dominant negative phenotype. Virology 178:327–330

Venkatesh LK, Mohammed S, Chinnadurai G (1990) Functional domains of the HIV-1 rev gene required for trans-regulation and subcellular localization. Virology 176:39–47

Weichselbraun I, Farrington GK, Rusche JR, Böhnlein E, Hauber J (1992) Definition of the human immunodeficiency virus type 1 Rev and human T-cell leukemia virus type I Rex protein activation domain by functional exchange. J Virol 66:2583–2587

Weis K (1998) Importins and exportins: how to get in and out of the nucleus [published erratum appears in Trends Biochem Sci 1998 Jul; 23(7):235]. Trends Biochem Sci 23:185–189

Wen W, Meinkoth JL, Tsien RY, Taylor SS (1995) Identification of a signal for rapid export of proteins from the nucleus. Cell 82:463–473

Wingfield PT, Stahl SJ, Payton MA, Venkatesan S, Misra M, Steven AJ (1991) HIV-1 rev expressed in recombinant *Escherichia coli*: purification, polymerization and conformational properties. Biochemistry 30:7527–7534

Wolff B, Cohen G, Hauber J, Meshcheryakova D, Rabeck C (1995) Nucleocytoplasmic transport of the Rev protein of human immunodeficiency virus type 1 is dependent on the activation domain of the protein. Exp Cell Res 217:31–41

Wolff B, Sanglier JJ, Wang Y (1997) Leptomycin B is an inhibitor of nuclear export: inhibition of nucleo-cytoplasmic translocation of the human immunodeficiency virus type 1 (HIV-1) Rev protein and Rev-dependent mRNA. Chem Biol 4:139–147

Zapp ML, Green MR (1989) Sequence-specific RNA binding by the HIV-1 Rev protein. Nature 342:714–716

Zapp ML, Hope TJ, Parslow TG, Green MR (1991) Oligomerization and RNA binding domains of the type 1 human immunodeficiency virus Rev protein: A dual function for an arginine-rich binding motif. Proc Natl Acad Sci USA 88:7734–7738

Zemmel RW, Kelley AC, Karn J, Butler PJG (1996) Flexible regions of RNA structure facilitate co-operative Rev assembly on the Rev-response element. J Mol Biol 258:763–777

Zhang G, Zapp ML, Yan G, Green MR (1996) Localization of HIV-1 RNA in mammalian nuclei. J Cell Biol 135:9–18

Zolotukhin AS, Felber BK (1999) Nucleoporins Nup98 and Nup214 participate in nuclear export of human immunodeficiency virus type 1 Rev. J Virol 73:120–127

Constitutive Transport Element-Mediated Nuclear Export

M.-L. Hammarskjöld

1	Introduction.	77
2	RNA Export in Simpler vs Complex Retroviruses	78
2.1	Retrovirus Replication Requires Export of Intron-Containing RNA	78
2.2	The Mason-Pfizer Monkey Virus (MPMV) Genome Contains a Constitutive Transport Element	80
2.3	Transport Elements in other Simpler Retroviruses	80
3	Structural and Mutational Analysis of the MPMV Constitutive Transport Element	81
4	The MPMV Constitutive Transport Element Promotes Export of RNA in *Xenopus* Oocytes	83
5	Identification of Cellular Proteins that Interact with the MPMV/SRV Constitutive Transport Element.	85
6	Constitutive Transport Element-Mediated RNA Export as a Model System for mRNA Export	86
7	Perspective and Future Directions	87
	References.	90

1 Introduction

The molecular biology of retrovirus replication and gene regulation has been intensely analyzed for many years. These studies have made it clear that retroviral expression is controlled not only at the transcriptional level, but also by posttranscriptional mechanisms that regulate the export of RNA from the host cell nucleus to the cytoplasm. This involves the action of proteins that actively shuttle between the nucleus and cytoplasm using specific nuclear import and export signals. Thus retroviruses have become important model systems for nucleo-cytoplasmic trafficking (for previous reviews, see Hammarskjold 1997; Cullen 1998; Pollard and Malim 1998).

The importance of posttranscriptional gene regulation in these viruses was first realized more than a decade ago in studies of two complex retroviruses, human

---

Myles H. Thaler Center for AIDS and Human Retrovirus Research, and the Department of Microbiology, University of Virginia, Charlottesville, VA 22908, USA

immunodeficiency virus (HIV) and human T-cell leukemia virus (HTLV). Several different laboratories demonstrated that the genomes of these viruses encode regulatory proteins (Rev in the case of HIV, Rex in the case of HTLV), that serve to promote the export of viral RNA from the nucleus to the cytoplasm (EMERMAN et al. 1989; FELBER et al. 1989; HAMMARSKJOLD et al. 1989; MALIM et al. 1989). At the time of these reports, very little was known about what regulates nuclear export in the eukaryotic cell. This sparked a great interest in the exact roles that Rev and Rex play in viral gene regulation and how these proteins interact with the cellular machinery.

Through subsequent studies it became clear that Rev and Rex serve to specifically regulate the nucleo-cytoplasmic export of unspliced and incompletely spliced RNAs. All of these RNAs retain complete introns. In addition, they contain structured RNA elements (RRE in HIV, RxRE in HTLV) to which the Rev/Rex proteins bind. Subsequent interaction of these proteins with cellular receptors involved in nuclear export promotes the nucleo-cytoplasmic transport of the viral RNAs. For a complete review of what is known to date about the Rev and Rex pathways and their roles as crucial model systems for elucidating cellular transport mechanisms, see the chapter by J. Hauber in this volume.

2 RNA Export in Simpler vs Complex Retroviruses

2.1 Retrovirus Replication Requires Export of Intron-Containing RNA

The retrovirus genome contains only one promoter, at the $5'$ end of the genome, and the only active transcription termination site is situated at the $3'$ end of the genome. Thus, retroviral transcription generates a single product. This "full-length" RNA serves as the genome that is packaged into progeny virus particles and is also used as mRNA to generate the internal viral capsid proteins and enzymes (i.e., the Gag/Pol proteins). The mRNAs for other viral proteins must be generated from this RNA by splicing of introns to remove the gag/pol region and other sequences to allow the translation of downstream exons (Fig. 1). In the case of complex retroviruses such as HIV, many different subgenomic mRNAs are generated, since several regulatory and accessory viral proteins are expressed in addition to the structural genes. Both single-spliced and multiply spliced RNAs are produced through removal of one or more introns. Since both the unspliced and single-spliced RNAs can undergo further splicing, both of these classes of RNA are thus by definition "intron-containing".

In the case of cellular genes, all introns are, as a general rule, removed before the RNA is exported from the nucleus, and incompletely spliced intron-containing RNAs are believed to be actively retained in the nucleus by specific proteins (CHANG and SHARP 1989; LEGRAIN and ROSBASH 1989). This may serve as an important mechanism to ensure that defective mRNAs do not accumulate in the

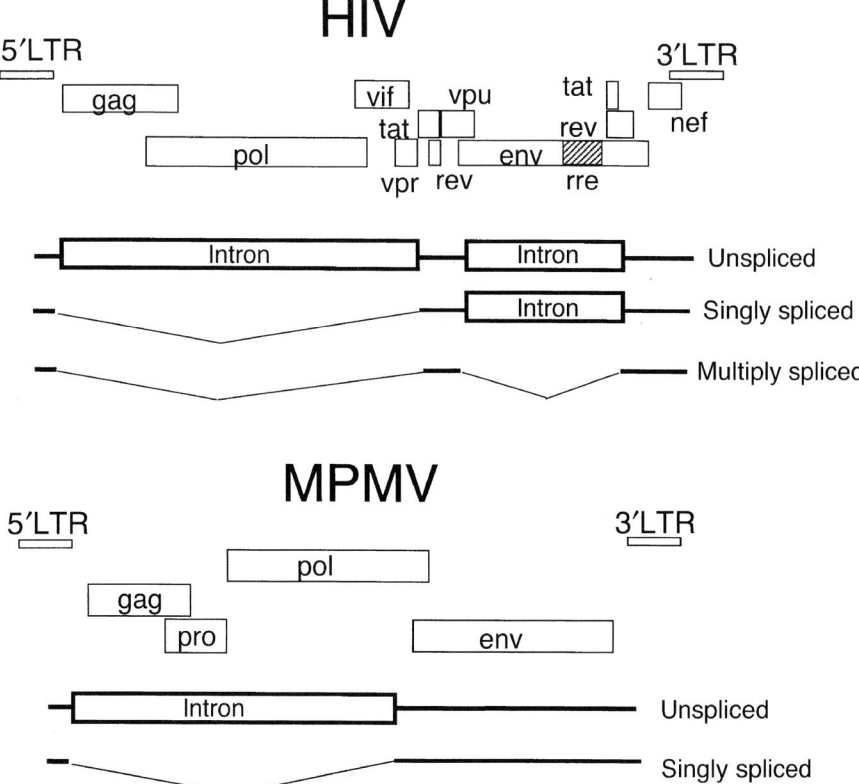

Fig. 1. Genomic organization of complex and simpler retroviruses and the viral RNAs that are expressed in infected cells. *Top*: The genome of human immundeficiency virus type 1 (HIV1) and the three principally different classes of RNAs that are generated and exported to the cytoplasm of infected cells. *Bottom*: The genome of Mason-Pfizer monkey virus (MPMV), a type D simpler retrovirus and the two different RNAs that are expressed in cells infected with this virus

cytoplasm, where they could be translated into aberrant proteins, leading to perturbation of cellular processes and thereby disease. The specific role of viral proteins such as Rev appears to be to override this restriction. Thus expression of Rev enables efficient export of intron-containing HIV RNAs that would otherwise be retained in the host cell nucleus (Lu et al. 1990; Olsen et al. 1992; Malim and Cullen 1993; Hammarskjold et al. 1994; Zhang et al. 1996; Fischer et al. 1999).

In the replication of simpler retroviruses, the mRNA that expresses the envelope protein is the only subgenomic message, since these viruses do not encode regulatory proteins (Fig. 1). However, since this RNA is always generated by splicing from the genomic RNA, the full-length message that expresses the Gag and Pol proteins is an intron-containing RNA. Thus, this RNA would be expected to be retained in the nucleus by cellular mechanisms. Since simpler retroviruses do not encode a Rev equivalent, it remained unclear for some time how the genomic RNA was able to overcome the expected cellular restriction.

2.2 The Mason-Pfizer Monkey Virus (MPMV) Genome Contains a Constitutive Transport Element

The first insight into this enigma came through the discovery of a *cis*-acting RNA export element in the genome of Mason-Pfizer Monkey virus, a simpler type D retrovirus. This element was initially identified because of its ability to efficiently substitute for Rev and the RRE in the expression of HIV proteins as well as in HIV replication (BRAY et al. 1994). Complementation of Rev/RRE function occurred only if the MPMV element was present in the normally Rev-dependent HIV RNAs in the "correct" orientation. Since the element, in contrast to the RRE, did not require a viral *trans*-activating protein to become active, it was given the name "constitutive transport element" (CTE).

The region necessary to complement Rev and RRE function was originally mapped to a 219-nucleotide fragment at the 3' end of the MPMV genome and was shown to be completely contained within an intragenic region between the *env* gene and the 3' long terminal repeat (LTR). Further mapping showed that the minimal element resides within a 154-nucleotide sequence (ERNST et al. 1997a). In spite of the fact that no viral functions had been previously mapped to the region containing the CTE, it was found to be highly conserved within the type D retroviruses, strengthening the hypothesis that the CTE might play a role in MPMV replication. Further studies showed that the CTE is indeed an essential element in replication of this virus (ERNST et al. 1997b). Deletion of the CTE completely abolished viral replication and, in line with the proposed role, CTE removal led to retention of genomic, unspliced RNA in the host cell nucleus. In contrast, the spliced env mRNA was exported to the cytoplasm even in the absence of the CTE and normal amounts of Env proteins were expressed. This confirmed the role of the CTE as a specific mediator of the export of intron-containing RNA.

2.3 Transport Elements in other Simpler Retroviruses

Several of the other simpler retroviruses have now been shown to contain sequences in their genomes that promote nuclear export. Another type D retrovirus, SRV1, also encodes a functional CTE that substitutes for the HIV Rev/RRE in expression from reporter constructs and replication (ZOLOTUKHIN et al. 1994). This is not surprising, since the MPMV and SRV1 CTE regions are more than 95% conserved at the nucleotide level. A similar element is present within an intracisternal-A particle element (TABERNERO et al. 1997), a retroviral transposon that was likely derived from a type D retrovirus.

An element that can function as a CTE has also been identified in the avian sarcoma/leukemia (ASV/ALV) retro viruses (OGERT et al. 1996; OGERT and BEEMON 1998). Although the exact role of this element in viral replication is still controversial (SIMPSON et al. 1997), this CTE can substitute for HIV Rev/RRE in subgenomic reporter constructs in avian cells. The element has been reported to function poorly in mammalian cells. The ASV/ALV CTE forms a secondary structure that is very different from that of the MPMV/SRV CTE, and there is no apparent sequence homology between the two elements (OGERT and BEEMON 1998; YANG and CULLEN 1999). Crm1 is not essential for function of the ASV/ALV CTE, as it is for Rev function, nor does this element appear to be recognized by the TAP protein, a cellular protein that binds specifically to the MPMV/SRV CTE (see below) (YANG and CULLEN 1999). Thus, this element may use yet a different pathway than the ones utilized by Rev/RRE and the MPMV/SRV CTE.

Unpublished experiments performed in our laboratory indicate that the Moloney murine leukemia virus genome may contain a CTE as well. This element maps to the *gag/pol* region and can substitute for Rev and the RRE in expression of an intron-containing RNA. No mutational or structural analysis has yet been performed on this potential element.

Although it would be expected that all retroviruses would encounter the same problem in exporting the full-length genomic RNA, it remains unclear whether a CTE is always required. Alternatively, it is possible that some of the retroviruses might overcome the cellular restriction to export by using weak splice signals that may result in less efficient retention of the RNA in the nucleus.

3 Structural and Mutational Analysis of the MPMV Constitutive Transport Element

When the MPMV CTE was discovered, it was immediately realized that this element is likely to interact directly with cellular proteins involved in nuclear export. Structural and mutational analysis of the MPMV and SRV1 CTE provided important clues as to the potential binding sites of such proteins (TABERNERO et al. 1996; ERNST et al. 1997a). The structural analysis demonstrated that the secondary structure of the MPMV CTE forms a long, single stem with two, internal single-stranded loops and a top loop (Fig. 2). Strikingly, the sequences of the two internal loops and surrounding areas of the stem are identical (shown in bold in Fig. 2), but "rotated" 180° relative to each other. The SRV CTE has been predicted to form an almost identical structure (TABERNERO et al. 1996). Mutations in and around either of the internal loop regions have been shown to completely abrogate both MPMV and SRV 1 CTE function in mammalian cells (TABERNERO et al. 1996; ERNST et al. 1997a; Fig. 2). In contrast, mutations in other areas of the stem are tolerated, provided that compensatory mutations are made on the "other" side of the stem to preserve the structure. This points to the importance of the internal loops for CTE function and strongly suggests that these loops provide the binding sites for a crucial cellular factor or factors. In the case of Rev, it has been shown that Rev multimerizes on the RRE and that Rev function requires the binding of at least two Rev molecules. The requirement for both internal loops for CTE function in mammalian cells might reflect a similar requirement for binding of at least two molecules of a cellular factor.

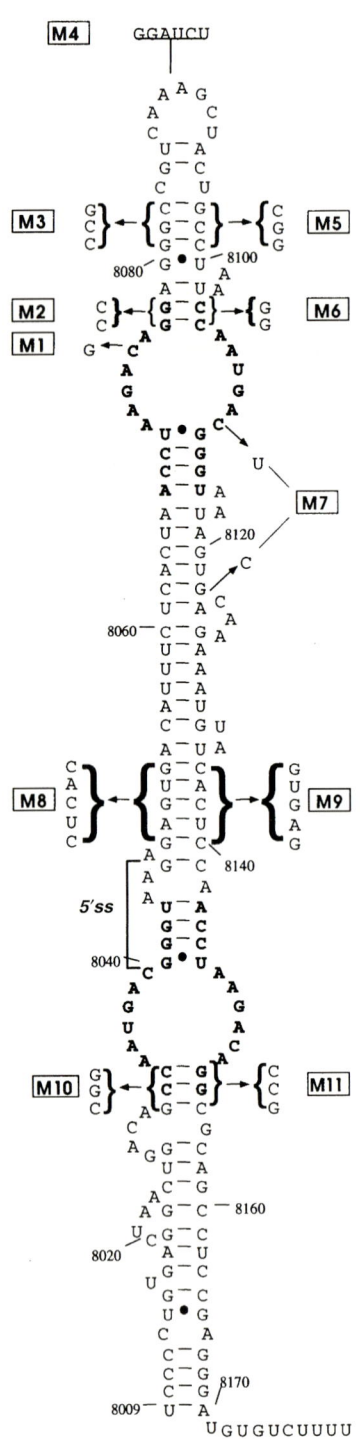

Mutant	Function in mammalian cells	Function in xenopus oocytes
WT	+	+
M1	−	+
M2	−	+
M4	−	+
M7	−	+
M9	−	+
M2/M6	−	+
M2/M11	−	−
M8/M9	+	+
M3/M5	+	ND
M10/M11	−	ND

Fig. 2. The secondary structure of the MPMV constitutive transport element (CTE) and the results of functional analysis of several CTE mutants in mammalian cells and *Xenopus* oocytes. The sequences of the two inner loops and surrounding areas are identical but rotated 180° relative to each other, as indicated in *bold*. In addition, the CTE contains a functional 5' splice site, as shown in the diagram

Although most studies to date have focused on the internal loops within the CTE, an insertion into the top loop (M4, Fig. 2) led to complete abrogation of CTE function in mammalian cells (Fig. 2; ERNST et al. 1997a). Thus it remains possible that the sequences in this loop are important for CTE function as well.

As indicated in Fig. 2, the CTE also contains a 5' splice site that has been shown to function in *Xenopus* oocytes (PASQUINELLI et al. 1997). The importance of this splice site is unknown, since no 3' splice site is present downstream of the CTE in the MPMV genome.

4 The MPMV Constitutive Transport Element Promotes Export of RNA in *Xenopus* Oocytes

The interest in the CTE was greatly enhanced by two studies that demonstrated that the CTE is able to promote nuclear export in *Xenopus* oocytes (PASQUINELLI et al. 1997; SAAVEDRA et al. 1997). The *Xenopus* system has provided an important model for RNA export for several years. Experiments in this system have shown that export of microinjected RNA is a saturable process that can be divided into different classes depending on competition (JARMOLOWSKI et al. 1994; FISCHER et al. 1995; NAKIELNY et al. 1997). It has been shown that tRNA, mRNA, ribosomal subunits and spliceosomal U snRNAs do not compete with each other for export in this system, and these classes of RNAs are thus likely to be exported using different pathways. However, it has been shown that 5S rRNA competes with export of many spliceosomal U snRNAs, indicating that these RNAs share one or more limiting cellular factors.

Several years ago, it was shown that export of RRE-containing RNA in *Xenopus* oocytes required co-injection of functional Rev protein, indicating that oocytes could provide a model system for Rev/RRE function (FISCHER et al. 1994, 1995). However, contrary to what might have been expected, Rev-mediated export did not compete with export of mRNA. Rather, competition was observed with 5S rRNA and U snRNAs. In subsequent studies in the oocyte system, Crm1 was identified as a cellular export receptor utilized by Rev (FORNEROD et al. 1997). Crm1 was shown to be a member of the β-family of transport receptors, a family that also includes several import receptors. Crm1 binds specifically to the domain of Rev that contains the nuclear export signal (NES), and a ternary complex is formed in the presence of RanGTP (ASKJAER et al. 1998). Ran is a small Ras-like protein that exists in both a GTP- and a GDP-bound form. Since the exchange factor for Ran (RCC1) is present mainly in the nucleus, whereas RanGAP localizes to the cytoplasm, GDP-bound Ran dominates in the cytoplasm, whereas RanGTP is predominantly nuclear. The steep RanGTP gradient that this generates across the nuclear membrane is believed to impart an important directionality to cellular transport. Import substrates are thought to bind to RanGDP in the cytoplasm and to be released by conversion to RanGTP in the nucleus. In contrast, export substrates form a complex with RanGTP in the nucleus and may be released by the action of RanGAP in the cytoplasm (for recent reviews, see GORLICH and KUTAY 1999; NAKIELNY and DREYFUSS 1999).

Rev function has been shown to be efficiently inhibited by the drug leptomycin B (LMB). This drug was initially identified as a cell cycle inhibitor in yeast and subsequently shown to inhibit HIV replication and Rev function in mammalian cells (WOLFF et al. 1997). LMB inhibits export of Rev also in *Xenopus* oocytes (FORNEROD et al. 1997), and it has now been shown that LMB specifically inhibits the ternary complex that can form in vitro between Rev, Crm1 and RanGTP (ASKJAER et al. 1998). The Crm1 protein has also been shown to contain a domain that interacts with several proteins that are part of the nuclear pore complexes (NPCs) (FORNEROD et al. 1997; ASKJAER et al. 1999). These large structures contain many different proteins (often known as nucleoporins or Nups) and have a molecular weight of 125 million Daltons in mammalian cells (for recent reviews, see MATTAJ and ENGLMEIER 1998; MOROIANU 1999). All the molecules that transport in and out of the nucleus have to pass through the nuclear pore, and this transport is believed to be an active process in the case of most proteins and all RNAs. Crm1 interacts specifically with the nucleoporin CAN/Nup214 as well as with other nucleoporins (FORNEROD et al. 1997).

The study by SAAVEDRA et al. (1997) investigated the export of the SRV CTE in oocytes in the context of a micro-injected intron-containing RNA. Although this RNA was not exported, the excised lariat containing the CTE reached the cytoplasm. Analysis of different CTE mutants demonstrated that a portion of the CTE containing one of the inner loops was sufficient for lariat export, demonstrating clear differences between this system and mammalian cells. In the study by PASQUINELLI et al. (1997), the export behavior of the CTE was studied using several different constructs. This study also demonstrated that the requirements for CTE-mediated export differ between oocytes and mammalian cells. Thus, several CTE mutants that were shown to be completely non-functional in mammalian cells were still able to promote export in the oocyte system (Fig. 2).

Based on the fact that the CTE and Rev/RRE pathways appeared interchangeable, it had been expected that the CTE would compete with Rev and 5S rRNA export. However, both of the initial studies of CTE-mediated export in the oocyte system convincingly demonstrated that this is not the case. Instead, the CTE was shown to specifically compete with export of mRNA substrates (PASQUINELLI et al. 1997; SAAVEDRA et al. 1997). At the time this was reported, very little was known about the cellular factors that mediate mRNA export. Thus, these papers promoted interest in the CTE as a model system and intensified the effort to identify cellular factors that interact with this element.

5 Identification of Cellular Proteins that Interact with the MPMV/SRV Constitutive Transport Element

To date there have been several reports of cellular factors in nuclear extracts from mammalian cells that bind to CTE RNA in vitro (PASQUINELLI et al. 1997; TANG et al. 1997; GRUTER et al. 1998). Two of these factors have been identified by microsequencing. The first factor to be identified was RNA helicase A, a protein that has also been proposed to be involved in transcriptional regulation (NAKAJIMA et al. 1997; TANG et al. 1997). A role for this protein in CTE function was suggested based on transient transfection experiments in which over-expression of RNA helicase A appeared to specifically promote CTE function (TANG et al. 1997). It was also reported that an excess of CTE RNA relocalized RNA helicase A to the cytoplasm. However, antibodies to RNA helicase A failed to inhibit CTE export in the oocyte system and experiments in our laboratory have not been able to confirm that RNA helicase A promotes CTE function in mammalian cells (PASQUINELLI et al. 1997). It was recently reported that RNA helicase A has effects also on the Rev/RRE-mediated pathway (LI et al. 1999).

The second CTE binding factor to be identified was a protein that was originally cloned and sequenced because of its association with the TIP protein from *Herpesvirus saimiri*, a simian herpesvirus (YOON et al. 1997). Because of this association, the protein was given the name TIP-associated protein (TAP). Although it was originally proposed that the TAP protein plays a role in *H. saimiri* gene regulation, this has not been further analyzed. TIP is a cytoplasmic protein that interacts with cellular tyrosine kinases, and there is no obvious link between this protein and nuclear export (YOON et al. 1997). The TAP protein was found to bind to CTE RNA in gel-shift assays and was also able to enhance the cytoplasmic accumulation in *Xenopus* oocytes of an excised lariat containing the CTE (GRUTER et al. 1998). Mutational analysis indicated a good correlation between the ability of the CTE to bind to TAP and export of the lariat. Since RNA containing the CTE is exported in oocytes even in the absence of added TAP, it is clear that the oocyte nucleus must contain endogenous proteins that are capable of promoting CTE function. Although it has been proposed that the *Xenopus* TAP equivalent may fulfill this role, this remains to be verified, since no *Xenopus* TAP gene has been identified to date.

Consistent with the proposed role as an RNA export receptor, TAP has recently been shown to be able to shuttle actively between the nucleus and cytoplasm of mammalian cells (BEAR et al. 1999; KANG and CULLEN 1999). The full-length TAP protein is 619 amino acids, and there seems to be general consensus that both the nuclear localization signal (NLS) and the region that binds to the CTE are present in the amino-terminal half of the protein (BEAR et al. 1999; BRAUN et al. 1999; KANG et al. 1999; KANG and CULLEN 1999; KATAHIRA et al. 1999). Two studies have demonstrated that the β-family receptor transportin acts as the import receptor for TAP (TRUANT et al. 1999; BACHI et al. 2000). In contrast, there is more controversy concerning the location of export signals in the TAP protein. Two studies have suggested the presence of a functional NES in the amino-terminal 100 amino acids of the protein(BEAR et al. 1999; BRAUN et al. 1999). In contrast, a recent study suggests that nucleo-cytoplasmic export of TAP in mammalian cells require sequences at the carboxy-terminal end of the protein (KANG and CULLEN 1999). This portion of TAP has been shown to interact with nucleoporins such as CAN/Nup214, p62 and Nup 98 (KATAHIRA et al. 1999; BACHI et al. 2000). Recent results from our laboratory also indicate that the TAP export signal resides within the carboxy-terminal half of the protein (GUZIK et al. 2001).

The interest in TAP as a potential mRNA export factor has also been sparked by the fact that this protein is a human orthologue of the yeast protein Mex67p. Mutation in MEX 67 was shown to lead to a rapid nuclear accumulation of polyA mRNA in yeast (SEGREF et al. 1997). It was reported that TAP could, to some extent, complement MEX function in yeast cells, but this was shown to require the presence of yet another human protein, p15 (also known as NXT1) (KATAHIRA et al. 1999). NXT1/p15 is a small protein with homology to NTF2 (BLACK et al. 1999; KATAHIRA et al. 1999), a protein that specifically binds RanGDP and plays a role in nuclear import (CORBETT and SILVER 1996; PASCHAL et al. 1996). In contrast, NXT1/p15 has been reported to be a RanGTP-binding factor (BLACK et al. 1999). NXT1/p15 binds directly to TAP in a region between the CTE-binding and the carboxy-terminal nucleoporin-binding domains (KATAHIRA et al. 1999). Whereas it is still unclear whether NXT1/p15 is essential for CTE function, recent experiments in our laboratory show that p15 greatly enhances TAP-mediated export in the context of a Rev M10/TAP fusion protein acting on the HIV RRE, where TAP provides NES function to an NES-deficient Rev protein (GUZIK et al. 2001).

Further support for a role for TAP in CTE RNA export has come from studies showing that over-expression of human TAP enables CTE-containing RNA to be exported in quail cells, where the CTE is not normally functional. However, preliminary data from our laboratory indicate that another cellular factor (SAM-68) is also capable of complementing CTE function in these cells, suggesting either that endogenous TAP can provide the necessary function in conjunction with SAM-68 or that TAP is not absolutely essential for CTE function. SAM (Src-associated in mitosis)-68 is a nuclear RNA-binding phospho-protein that serves as a major substrate for tyrosine phosphorylation by the cellular Src protein during mitosis (FUMAGALLI et al. 1994; TAYLOR and SHALLOWAY 1994; LOCK et al. 1996). This protein may thus provide a link between signal transduction and nuclear export pathways. SAM-68 was recently proposed to play a role also in Rev-mediated RNA export and to bind directly to the RRE (REDDY et al. 1999). It is still unclear whether SAM-68 binds directly to the CTE.

6 Constitutive Transport Element-Mediated RNA Export as a Model System for mRNA Export

One of the most important outstanding questions is how CTE export relates to export of cellular mRNA. Although TAP has been proposed as an important cellular factor in this respect, it is clear that most mRNAs do not contain specific signals similar to the CTE. If TAP is indeed a general mRNA export receptor, the mechanism for binding of this factor to RNA will thus have to be different from the one for CTE binding. It has been reported that TAP binds to the CTE with a 3,000 times higher affinity compared to binding to mRNA lacking the CTE (BACHI et al. 2000). One possibility would be for TAP to interact with mRNA through a protein (e.g., an hnRNP protein) rather than through direct RNA binding. Alternatively, the affinity of TAP for RNA could be increased through the action of other proteins.

To date, the data implying TAP as a general mRNA export factor come mainly from studies in *S. cerevisiae* (KATAHIRA et al. 1999). One major difference between this organism and metazoans is that most metazoan mRNAs undergo splicing, whereas most *S. cerevisiae* genes lack introns. A recent report by Reed and co-workers suggests that splicing generates specific nucleoprotein complexes that target mRNA for export in metazoans (LUO and REED 1999). Thus it was observed that RNAs derived from cDNAs and thus lacking introns ("Δi-RNAs") were poor substrates for export in *Xenopus* oocytes compared to RNAs that were allowed to undergo splicing in this system. Also, RNA that was spliced before export did not compete with the export of Δi-RNA. These results complicate the interpretation of the previous CTE experiments in *Xenopus* oocytes, in which the competition experiments were performed exclusively with Δi-RNA. In view of this, it seems possible that TAP is not involved in the export of mRNA that undergoes complete splicing in mammalian cells, but rather is specifically involved in export of intron-containing and/or RNA expressed from genes lacking introns.

7 Perspective and Future Directions

As evidenced by this review, numerous studies concerning CTE function and how this relates to Rev/RRE function have been published since the CTE was first discovered. The most surprising finding, documented in several studies, is undoubtedly that the CTE and the Rev/RRE appear to work through cellular pathways that are largely non-overlapping. Since these elements apparently perform the same function and appear completely interchangeable, it had originally been hypothesized that the CTE might utilize a Rev-like cellular factor. However, although there is still controversy concerning CTE cellular cofactors, all the available data suggest that the CTE utilizes a pathway that is independent of Crm1, the cellular cofactor shown to be of crucial importance for Rev export. This points to the complexity of cellular export pathways. It is clear that our understanding of these processes is still in its infancy, despite the remarkable progress that has been made in the last few years.

One of the most important goals for future research will be to elucidate how RNA export relates to other cellular processes, such as transcription, splicing and translation. Recent studies suggest that the CTD of RNA polymerase II plays a direct role in directing splicing and other factors to the RNA during synthesis. Thus, the eventual fate of an mRNA may be determined already at this level (for a recent review, see CARMO-FONSECA et al. 1999). At least one published report suggests that Rev has to be recruited to the RNA during synthesis (IACAMPO and COCHRANE 1996). It therefore seems possible that the same will turn out to be true for the cellular factors that promote CTE-mediated export. Based on results that suggested that the CTE needed to be close to a poly A signal to function efficiently, a functional interaction between the CTE and the polyadenylation machinery has also been proposed (RIZVI et al. 1997). However, previous studies in our laboratory indicated that the CTE can function efficiently even when it is far away from a polyA signal (ERNST et al. 1997b). A recent report from another laboratory seems to confirm this (WODRICH et al. 2000). One reason for these seemingly conflicting results might be the presence of a functional 5′ splice site in the CTE (Fig. 2). By removing the CTE further away from the polyA signal, RIZVI et al. (1997) may have inadvertently created constructs in which the CTE was removed by splicing between the 5′ splice site and a cryptic splice site in downstream sequences. Consistent with this, the recent study by WODRICH et al. (2000) showed that the CTE did not function when placed upstream of an intron.

As far as the relationship between the export of intron-containing RNA and the splicing machinery is concerned, it was demonstrated several years ago that splice sites appear to play an important role in Rev function (CHANG and SHARP 1989; LU et al. 1990; HAMMARSKJOLD et al. 1994). It has also been demonstrated that Rev can act on an RNA that is also recognized by splicing factors (LU et al. 1990). Unpublished data from our laboratory have shown that the same holds true for CTE function. It thus seems likely that the CTE somehow interacts with components of the splicing machinery to "rescue" intron-containing RNA and enable its export. Increasing evidence is mounting that novel protein complexes are formed on RNA during the process of splicing, and these complexes promote RNA export once splicing is completed (LUO and REED 1999; LE HIR et al. 2000). It will thus be of great future interest to determine if any of the factors that are present in these complexes also participate in the Rev/RRE- or CTE-mediated pathways.

Although it is clear that both the Rev/RRE and the CTE function to directly promote nuclear export, as evidenced by retention of intron-containing viral RNA in the nucleus when either of these pathways are perturbed, there also seem to be some effects at the cytoplasmic level. For example, it has been observed that effects on protein expression in the absence of the Rev/RRE or CTE are more dramatic with some constructs than might be expected solely based on differences in RNA levels (LAWRENCE et al. 1991; D'AGOSTINO et al. 1992). Also, recent experiments in our laboratory suggest that the effect of SAM-68 on CTE function (see above) may be, to a large extent, at the level of cytoplasmic utilization rather than at the level of export. In view of this, it will be of importance to determine whether Rev/RRE and/or the CTE may also affect the nature of the protein complexes that remain on the RNA after the export process is completed. The nature of such complexes could indirectly or directly affect translation of proteins from the exported RNA. In fact, other members of the STAR (signal transduction and activator of RNA) family of proteins, to which SAM-68 belongs, have already been shown to associate with specific RNAs to regulate translation (JAN et al. 1999; SACCOMANNO et al. 1999).

The effects of SAM-68 on RNA export pathways may also point to yet another link between signal transduction and regulation of RNA at the post-transcriptional level. Recent data from several laboratories have shown that signal transduction affects phosphorylation of several splicing factors, with secondary effects on both constitutive and alternative splicing (PRASAD et al. 1999; STOJDL and BELL 1999; YEAKLEY et al. 1999). Perhaps even more significant is the recent identification of PHAX, a protein involved in U snRNA export (OHNO et al. 2000). PHAX is phosphorylated in the nucleus and then exported with RNA to the cytoplasm, where it is dephosphorylated. PHAX phosphorylation is essential for export complex assembly, while its dephosphorylation was reported to cause export complex disassembly. The details of the mechanisms by which SAM-68, PHAX and other phosphorylated RNA-interacting proteins regulate posttranscriptional events, and the effects this has on gene expression, will be an important area of investigation for the future.

Since viruses usually adapt cellular mechanism to solve their expression problems, it seems reasonable to hypothesize that the CTE is mimicking a process that is also of importance in cellular gene regulation. In view of this, it will be of obvious interest to determine whether there are true cellular counterparts to the MPMV CTE. Despite efforts in our laboratory and in others to identify such potential cellular CTEs, none have been found to date. It is tempting to speculate that such sequences might be present in cellular mRNAs that reach the cytoplasm with retained introns. While there are only few examples of this in the literature, such RNAs have in some instances been shown to generate alternative proteins that may serve a regulatory function. One example of this is the Id family of genes, which encode negative regulators of basic helix-loop-helix transcription factors. At least two members of this family (Id1 and Id3) have been shown to be capable of generating alternative proteins with unique carboxy-terminal domains by retention of an intron that is removed from the RNA under other conditions (SPRINGHORN et al. 1994; DEED et al. 1996; NEHLIN et al. 1997). These isoforms have been proposed to modulate Id function during differentiation. Another example of intron retention occurs in many different types of cancer, in which intron-containing forms of CD44 can be detected in malignant cells (YOSHIDA et al. 1995; BOLODEOKU et al. 1996; STICKELER et al. 1997; GOODISON et al. 1998; GORHAM et al. 1999). Further studies of the regulation of export of intron-containing RNAs and the search for cellular CTEs may help discover hitherto unknown mechanisms that are of importance for cellular gene regulation in normal and transformed cells.

Acknowledgements. I would to like to thank David Rekosh for our long-standing collaboration in the area of Rev/RRE and CTE function and for many fruitful and challenging discussions. The contributions of our many students and post-docs throughout the years are also gratefully acknowledged. I would also like to acknowledge the support of the University of Virginia, where I am the Charles H. Ross Jr. Professor and receive support from the Charles H. Ross Jr. and the Myles H. Thaler Endowments. Current work in the author's laboratory is supported by NIH grant R01 AI34721.

References

Askjaer P, Bachi A, Wilm M, Bischoff FR, Weeks DL, Ogniewski V, Ohno M, Niehrs C, Kjems J, Mattaj IW, et al. (1999) RanGTP-regulated interactions of CRM1 with nucleoporins and a shuttling DEAD-box helicase. Mol Cell Biol 19:6276–6285

Askjaer P, Jensen TH, Nilsson J, Englmeier L, Kjems J (1998) The specificity of the CRM1-Rev nuclear export signal interaction is mediated by RanGTP. J Biol Chem 273:33414–33422

Bachi A, Braun IC, Rodrigues JP, Pante N, Ribbeck K, von Kobbe C, Kutay U, Wilm M, Gorlich D, Carmo-Fonseca M, et al. (2000) The C-terminal domain of TAP interacts with the nuclear pore complex and promotes export of specific CTE-bearing RNA substrates. RNA 6:136–158

Bear J, Tan W, Zolotukhin AS, Tabernero C, Hudson EA, Felber BK (1999) Identification of novel import and export signals of human TAP, the protein that binds to the constitutive transport element of the type D retrovirus mRNAs. Mol Cell Biol 19:6306–6317

Black BE, Levesque L, Holaska JM, Wood TC, Paschal BM (1999) Identification of an NTF2-related factor that binds Ran-GTP and regulates nuclear protein export. Mol Cell Biol 19:8616–8624

Bolodeoku J, Yoshida K, Sugino T, Goodison S, Tarin D (1996) Accumulation of immature intron-containing CD44 gene transcripts in breast cancer tissues. Mol Diagn 1:175–181

Braun IC, Rohrbach E, Schmitt C, Izaurralde E (1999) TAP binds to the constitutive transport element (CTE) through a novel RNA-binding motif that is sufficient to promote CTE-dependent RNA export from the nucleus. EMBO J 18:1953–1965

Bray M, Prasad S, Dubay JW, Hunter E, Jeang KT, Rekosh D, Hammarskjöld ML (1994) A small element from the Mason-Pfizer monkey virus genome makes human immunodeficiency virus type 1 expression and replication Rev-independent. Proc Natl Acad Sci USA 91:1256–1260

Carmo-Fonseca M, Custodio N, Calado A (1999) Intranuclear trafficking of messenger RNA. Crit Rev Eukaryot Gene Expr 9:213–219

Chang DD, Sharp PA (1989) Regulation by HIV Rev depends upon recognition of splice sites. Cell 59:789–795

Corbett AH, Silver PA (1996) The NTF2 gene encodes an essential, highly conserved protein that functions in nuclear transport in vivo. J Biol Chem 271:18477–18484

Cullen BR (1998) Retroviruses as model systems for the study of nuclear RNA export pathways. Virology 249:203–210

D'Agostino DM, Felber BK, Harrison JE, Pavlakis GN (1992) The Rev protein of human immunodeficiency virus type 1 promotes polysomal association and translation of gag/pol and vpu/env mRNAs. Mol Cell Biol 12:1375–1386

Deed RW, Jasiok M, Norton JD (1996) Attenuated function of a variant form of the helix-loop-helix protein, Id-3, generated by an alternative splicing mechanism. FEBS Lett 393:113–116

Emerman M, Vazeux R, Peden K (1989) The rev gene product of the human immunodeficiency virus affects envelope-specific RNA localization. Cell 57:1155–1165

Ernst R, Bray M, Rekosh D, Hammarskjöld M-L (1997a) A structured retroviral RNA element that mediates nucleocytoplasmic export of intron-containing RNA. Mol Cell Biol 17:135–144

Ernst RK, Bray M, Rekosh D, Hammarskjold ML (1997b) Secondary structure and mutational analysis of the Mason-Pfizer monkey virus RNA constitutive transport element. RNA 3:210–222

Felber BK, Hadzopoulou-Cladaras M, Cladaras C, Copeland T, Pavlakis GN (1989) Rev protein of human immunodeficiency virus type 1 affects the stability and transport of the viral mRNA. Proc Natl Acad Sci USA 86:1496–1499

Fischer U, Huber J, Boelens WC, Mattaj IW, Luhrmann R (1995) The HIV-1 Rev activation domain is a nuclear export signal that accesses an export pathway used by specific cellular RNAs. Cell 82:475–483

Fischer U, Meyer S, Teufel M, Heckel C, Luhrmann R, Rautmann G (1994) Evidence that HIV-1 Rev directly promotes the nuclear export of unspliced RNA. EMBO J 13:4105–4112

Fischer U, Pollard VW, Luhrmann R, Teufel M, Michael MW, Dreyfuss G, Malim MH (1999) Rev-mediated nuclear export of RNA is dominant over nuclear retention and is coupled to the Ran-GTPase cycle. Nucleic Acids Res 27:4128–4134

Fornerod M, Ohno M, Yoshida M, Mattaj IW (1997) CRM1 is an export receptor for leucine-rich nuclear export signals. Cell 90:1051–1060

Fornerod M, van Deursen J, van Baal S, Reynolds A, Davis D, Murti KG, Fransen J, Grosveld G (1997) The human homologue of yeast CRM1 is in a dynamic subcomplex with CAN/Nup214 and a novel nuclear pore component Nup88. EMBO J 16:807–816

Fumagalli S, Totty NF, Hsuan JJ, Courtneidge SA (1994) A target for Src in mitosis. Nature 368:871–874
Goodison S, Yoshida K, Churchman M, Tarin D (1998) Multiple intron retention occurs in tumor cell CD44 mRNA processing. Am J Pathol 153:1221–1228
Gorham H, Woodman A, Goodison S, Marsh J, Charnock M, Manek S, Sugino T, Tarin D (1999) CD44 expression in cervical intraepithelial neoplasia (CIN) and carcinoma. Mol Diagn 4:45–56
Gorlich D, Kutay U (1999) Transport between the cell nucleus and the cytoplasm. Annu Rev Cell Dev Biol 15:607–660
Gruter P, Tabernero C, von KC, Schmitt C, Saavedra C, Bachi A, Wilm M, Felber BK, Izaurralde E (1998) TAP, the human homolog of Mex67p, mediates CTE-dependent RNA export from the nucleus. Mol Cell 1:649–659
Guzik B, Prasad S, Levesque L, Black BE, Paschal BM, Rekosh D, Hammarskjöld M-L (2001) NXT1/p15 is a crucial cellular cofactor in TAP mediated export of intron-containing RNA in mammalian cells. Mol Cell Biol (in press)
Hammarskjold M-L (1997) Regulation of retroviral RNA export. Sem in Cell Dev Bio 8:83–90
Hammarskjold M-L, Heimer J, Hammarskjold B, Sangwan I, Albert L, Rekosh D (1989) Regulation of human immunodeficiency virus env expression by the rev gene product. J Virol 63:1959–1966
Hammarskjold ML, Li H, Rekosh D, Prasad S (1994) Human immunodeficiency virus env expression becomes Rev-independent if the env region is not defined as an intron. J Virol 68:951–958
Iacampo S, Cochrane A (1996) Human immunodeficiency virus type 1 Rev function requires continued synthesis of its target mRNA. J Virol 70:8332–8339
Jan E, Motzny CK, Graves LE, Goodwin EB (1999) The STAR protein, GLD-1, is a translational regulator of sexual identity in *Caenorhabditis elegans*. EMBO J 18:258–269
Jarmolowski A, Boelens WC, Izaurralde E, Mattaj IW (1994) Nuclear export of different classes of RNA is mediated by specific factors. J Cell Biol 124:627–635
Kang Y, Bogerd HP, Yang J, Cullen BR (1999) Analysis of the RNA binding specificity of the human tap protein, a constitutive transport element-specific nuclear RNA export factor. Virology 262:200–209
Kang Y, Cullen BR (1999) The human Tap protein is a nuclear mRNA export factor that contains novel RNA-binding and nucleocytoplasmic transport sequences. Genes Dev 13:1126–1139
Katahira J, Strasser K, Podtelejnikov A, Mann M, Jung JU, Hurt E (1999) The Mex67p-mediated nuclear mRNA export pathway is conserved from yeast to human. EMBO J 18:2593–2609
Lawrence JB, Cochrane AW, Johnson CV, Perkins A, Rosen CA (1991) The HIV-1 Rev protein: a model system for coupled RNA transport and translation. New Biol 3:1220–1232
Le Hir H, Moore MJ, Maquat LE (2000) Pre-mRNA splicing alters mRNP composition: evidence for stable association of proteins at exon-exon junctions. Genes Dev 14:1098–1108
Legrain P, Rosbash M (1989) Some cis- and trans-acting mutants for splicing target pre-mRNA to the cytoplasm. Cell 57:573–583
Li J, Tang H, Mullen TM, Westberg C, Reddy TR, Rose DW, Wong-Staal F (1999) A role for RNA helicase A in post-transcriptional regulation of HIV type 1. Proc Natl Acad Sci USA 96:709–714
Lock P, Fumagalli S, Polakis P, McCormick F, Courtneidge SA (1996) The human p62 cDNA encodes Sam68 and not the RasGAP-associated p62 protein. Cell 84:23–24
Lu XB, Heimer J, Rekosh D, Hammarskjold ML (1990) U1 small nuclear RNA plays a direct role in the formation of a rev-regulated human immunodeficiency virus env mRNA that remains unspliced. Proc Natl Acad Sci USA 87:7598–7602
Luo MJ, Reed R (1999) Splicing is required for rapid and efficient mRNA export in metazoans. Proc Natl Acad Sci USA 96:14937–14942
Malim MH, Cullen BR (1993) Rev and the fate of pre-mRNA in the nucleus: implications for the regulation of RNA processing in eukaryotes. Mol Cell Biol 13:6180–6189
Malim MH, Hauber J, Le SV, Maizel JV, Cullen BR (1989) The HIV-1 rev trans-activator acts through a structured target sequence to activate nuclear export of unspliced viral mRNA. Nature 338:254–257
Mattaj IW, Englmeier L (1998) Nucleocytoplasmic transport: the soluble phase. Annu Rev Biochem 67:265–306
Moroianu J (1999) Nuclear import and export: transport factors, mechanisms and regulation. Crit Rev Eukaryot Gene Expr 9:89–106
Nakajima T, Uchida C, Anderson SF, Lee CG, Hurwitz J, Parvin JD, Montminy M (1997) RNA helicase A mediates association of CBP with RNA polymerase II. Cell 90:1107–1112
Nakielny S, Dreyfuss G (1999) Transport of proteins and RNAs in and out of the nucleus. Cell 99:677–690
Nakielny S, Fischer U, Michael WM, Dreyfuss G (1997) RNA transport. Annu Rev Neurosci 20:269–301

Nehlin JO, Hara E, Kuo WL, Collins C, Campisi J (1997) Genomic organization, sequence, and chromosomal localization of the human helix-loop-helix Id1 gene. Biochem Biophys Res Commun 231:628–634

Ogert RA, Beemon KL (1998) Mutational analysis of the rous sarcoma virus DR posttranscriptional control element. J Virol 72:3407–3411

Ogert RA, Lee LH, Beemon KL (1996) Avian retroviral RNA element promotes unspliced RNA accumulation in the cytoplasm. J Virol 70:3834–3843

Ohno M, Segref A, Bachi A, Wilm M, Mattaj IW (2000) PHAX, a mediator of U snRNA nuclear export whose activity is regulated by phosphorylation. Cell 101:187–198

Olsen HS, Cochrane AW, Rosen C (1992) Interaction of cellular factors with intragenic *cis*-acting repressive sequences within the HIV genome. Virology 191:709–715

Paschal BM, Delphin C, Gerace L (1996) Nucleotide-specific interaction of Ran/TC4 with nuclear transport factors NTF2 and p97. Proc Natl Acad Sci USA 93:7679–7683

Pasquinelli AE, Ernst RK, Lund E, Grimm C, Zapp ML, Rekosh D, Hammarskjold ML, Dahlberg JE (1997) The constitutive transport element (CTE) of Mason-Pfizer monkey virus (MPMV) accesses a cellular mRNA export pathway. EMBO J 16:7500–7510

Pollard VW, Malim MH (1998) The HIV-1 Rev protein. Annu Rev Microbiol 52:491–532

Prasad J, Colwill K, Pawson T, Manley JL (1999) The protein kinase Clk/Sty directly modulates SR protein activity: both hyper- and hypophosphorylation inhibit splicing. Mol Cell Biol 19:6991–7000

Reddy TR, Xu W, Mau JK, Goodwin CD, Suhasini M, Tang H, Frimpong K, Rose DW, Wong-Staal F (1999) Inhibition of HIV replication by dominant negative mutants of Sam68, a functional homolog of HIV-1 Rev. Nat Med 5:635–642

Rizvi TA, Schmidt RD, Lew KA (1997) Mason-Pfizer monkey virus (MPMV) constitutive transport element (CTE) functions in a position-dependent manner. Virology 236:118–129

Saavedra C, Felber B, Izaurralde E (1997) The simian retrovirus-1 constitutive transport element, unlike the HIV-1 RRE, uses factors required for cellular mRNA export. Curr Biol 7:619–628

Saccomanno L, Loushin C, Jan E, Punkay E, Artzt K, Goodwin EB (1999) The STAR protein QKI-6 is a translational repressor. Proc Natl Acad Sci USA 96:12605–12610

Segref A, Sharma K, Doye V, Hellwig A, Huber J, Luhrmann R, Hurt E (1997) Mex67p, a novel factor for nuclear mRNA export, binds to both poly(A)+ RNA and nuclear pores. EMBO J 16:3256–3271

Simpson SB, Zhang L, Craven RC, Stoltzfus CM (1997) Rous sarcoma virus direct repeat cis elements exert effects at several points in the virus life cycle. J Virol 71:9150–9156

Springhorn JP, Singh K, Kelly RA, Smith TW (1994) Posttranscriptional regulation of Id1 activity in cardiac muscle. Alternative splicing of novel Id1 transcript permits homodimerization. J Biol Chem 269:5132–5136

Stickeler E, Mobus VJ, Kieback DG, Kohlberger P, Runnebaum IB, Kreienberg R (1997) Intron 9 retention in gene transcripts suggests involvement of CD44 in the tumorigenesis of ovarian cancer. Anticancer Res 17:4395–4398

Stojdl DF, Bell JC (1999) SR protein kinases: the splice of life. Biochem Cell Biol 77:293–298

Tabernero C, Zolotukhin AS, Bear J, Schneider R, Karsenty G, Felber BK (1997) Identification of an RNA sequence within an intracisternal-A particle element able to replace Rev-mediated posttranscriptional regulation of human immunodeficiency virus type 1. J Virol 71:95–101

Tabernero C, Zolotukhin AS, Valentin A, Pavlakis GN, Felber BK (1996) The posttranscriptional control element of the simian retrovirus type 1 forms an extensive RNA secondary structure necessary for its function. J Virol 70:5998–6011

Tang H, Gaietta GM, Fischer WH, Ellisman MH, Wong-Staal F (1997) A cellular cofactor for the constitutive transport element of type D retrovirus. Science 276:1412–1415

Tang H, Xu Y, Wong-Staal F (1997) Identification and purification of cellular proteins that specifically interact with the RNA constitutive transport elements from retrovirus D. Virology 228:333–339

Taylor SJ, Shalloway D (1994) An RNA-binding protein associated with Src through its SH2 and SH3 domains in mitosis. Nature 368:867–871

Truant R, Kang Y, Cullen BR (1999) The human tap nuclear RNA export factor contains a novel transportin-dependent nuclear localization signal that lacks nuclear export signal function. J Biol Chem 274:32167–32171

Wodrich H, Schambach A, Krausslich HG (2000) Multiple copies of the Mason-Pfizer monkey virus constitutive RNA transport element lead to enhanced HIV-1 Gag expression in a context-dependent manner. Nucleic Acids Res 28:901–910

Wolff B, Sanglier JJ, Wang Y (1997) Leptomycin B is an inhibitor of nuclear export: inhibition of nucleo-cytoplasmic translocation of the human immunodeficiency virus type 1 (HIV-1) Rev protein and Rev-dependent mRNA. Chem Biol 4:139–147

Yang J, Cullen BR (1999) Structural and functional analysis of the avian leukemia virus constitutive transport element. RNA 5:1645–1655

Yeakley JM, Tronchere H, Olesen J, Dyck JA, Wang HY, Fu XD (1999) Phosphorylation regulates in vivo interaction and molecular targeting of serine/arginine-rich pre-mRNA splicing factors. J Cell Biol 145:447–455

Yoon DW, Lee H, Seol W, DeMaria M, Rosenzweig M, Jung JU (1997) Tap: a novel cellular protein that interacts with tip of herpesvirus saimiri and induces lymphocyte aggregation. Immunity 6:571–582

Yoshida K, Bolodeoku J, Sugino T, Goodison S, Matsumura Y, Warren BF, Toge T, Tahara E, Tarin D (1995) Abnormal retention of intron 9 in CD44 gene transcripts in human gastrointestinal tumors. Cancer Res 55:4273–4277

Zhang G, Zapp ML, Yan G, Green MR (1996) Localization of HIV-1 RNA in mammalian nuclei. J Cell Biol 135:9–18

Zolotukhin AS, Valentin A, Pavlakis GN, Felber BK (1994) Continuous propagation of RRE(−) and Rev(−)RRE(−) human immunodeficiency virus type 1 molecular clones containing a cis-acting element of simian retrovirus type 1 in human peripheral blood lymphocytes. J Virol 68:7944–7952

Nuclear Pore Complex Architecture and Functional Dynamics

B. Fahrenkrog[1], D. Stoffler[1,2], and U. Aebi[1]

1 Introduction 95
2 Identification, Three-Dimensional Location and Functional Characterization of Nucleoporins 96
3 The Nuclear Pore Complex Is a Dynamic Structure 99
4 Interactions Between Nucleoporins and Viral Proteins 102
5 Nuclear Pore Complex Architecture 104
6 Functional States of the Nuclear Pore Complex 105
7 Controversial Nuclear Pore Complex Components: the "Central Plug" 109
8 Nuclear Pore Complex Disassembly and Reassembly 110
9 Conclusions 111
References 111

1 Introduction

The vertebrate nuclear pore complex (NPC) is a ~125-MDa supramolecular assembly, embedded in the double membrane of the nuclear envelope (NE). The NPC enables passive diffusion of ions and small molecules and mediated, signal-dependent, bidirectional nucleocytoplasmic transport of proteins, RNAs and RNP particles (reviewed in Izaurralde and Adam 1998; Ohno et al. 1998; Mattaj and Englmeier 1998; Adam 1999). Extensive electron microscopic (EM) analyses of isolated amphibian NEs have revealed a consensus model of the three-dimensional (3-D) architecture of the NPC (reviewed in Panté and Aebi 1996a; Stoffler et al. 1999a). Accordingly, the NPC consists of a ~55-MDa central framework (also termed "spoke complex") that is sandwiched between a ~32-MDa cytoplasmic ring and a ~21-MDa nuclear ring (Fig. 1). The cytoplasmic ring is decorated by eight ~50-nm-long kinky fibrils, whereas a "basket" (or "fishtrap") tops the nuclear ring, which is assembled from eight ~50–100-nm-long fibrils that join distally to form a

---

[1] Biozentrum, M.E. Müller Institute for Structural Biology, University of Basel, 4056 Basel, Switzerland
[2] Present address: The Scripps Research Institute, La Jolla, CA 92037, USA
B. Fahrenkrog and D. Stoffler have contributed equally to this review

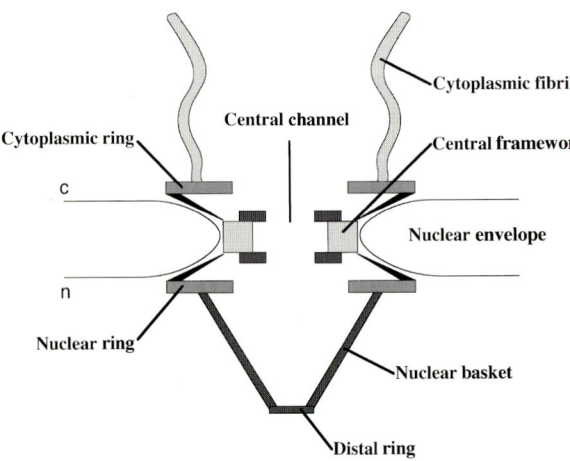

Fig. 1. The nuclear pore complex (NPC) architecture. *c*, cytoplasm; *n*, nucleus

30–50-nm diameter distal ring. The ring-like central framework embraces a central pore or channel that mediates the signal-dependent transport of molecules in and out of the nucleus. Ions as well as small molecules diffuse passively between the cytoplasm and the nucleus, possibly through eight peripheral channels that perforate the central framework. Frequently, the central channel appears plugged by a "particle" of variable size and morphology whose definite structure and function remain to be established (see below).

In this chapter, we review recent findings concerning: (1) the identification, 3-D location, and functional characterization of individual NPC constituents (i.e., nucleoporins) including their potential role in viral transport; (2) the 3-D architecture and structure-based functional dynamics of the NPC and some of its components; and (3) pathways and mechanisms involved in NPC disassembly and reassembly.

2 Identification, Three-Dimensional Location and Functional Characterization of Nucleoporins

Based on its molecular mass of ~125 MDa, the 822 symmetry of the central framework and the eight-fold symmetry of its cytoplasmic and nuclear periphery, it is assumed that the vertebrate NPC is composed of ~100 different proteins, called nucleoporins (Nups; reviewed in PANTÉ and AEBI 1996a; STOFFLER et al. 1999a). The smaller, ~60-MDa yeast NPC is assumed to be composed of about 30–50 different nucleoporins. Recent EM studies, however, have demonstrated that, although the yeast NPC appears to be ~15% smaller in its linear dimensions, the overall 3-D architecture of the NPC seems to be highly conserved from yeast to higher eukaryotes (FAHRENKROG et al. 1998; YANG et al. 1998). As of today, ~20 vertebrate and ~30 yeast nucleoporins have been identified and characterized (see STOFFLER et al. 1999a; see also ROUT et al. 2000).

A common feature of a number of nucleoporins are FG repeat sequence motifs within their amino acid sequence. Evidence has been provided that these sequence repeat motifs play a functional role in nucleocytoplasmic transport (see REXACH and BLOBEL 1995). Whereas in vitro interactions between FG nucleoporins and transport factors (e.g., as evaluated by blot overlays) appear to be rather unspecific (see RADU et al. 1995), in vivo assays (*Xenopus* egg extracts) could demonstrate more specific interactions between nucleoporins and transport factors (IOVINE and WENTE 1997; MARELLI et al. 1998; SHAH and FORBES 1998; SHAH et al. 1998; STOCHAJ et al. 1998; SEEDORF et al. 1999). In this context, the FG repeats might function as docking sites for cargo complexes. However, not all FG repeat-type docking sites might be obligatory binding sites for a given cargo complex, rather they may serve to increase the yield of docking cargo. Additionally, the FG repeats might not only function as docking sites but also as "parking" sites to keep cargo in a waiting position at the NPC for subsequent passage through the central pore.

An in vivo interaction between Nup153 and the import receptor importin-β has been demonstrated in *Xenopus* egg extracts (SHAH et al. 1998). Since Nup153 is located near or at the distal ring of the nuclear basket (PANTÉ et al. 1994), it is conceivable that Nup153 is the termination site for nuclear protein import before the cargo complex is dissociated and the cargo released into the nucleus (SHAH et al. 1998). Tpr, a constituent of the distal end of the nuclear basket and/or the intranuclear filaments associated with the nuclear basket at the level of its distal ring, also interacts in vivo with the import receptor dimer importin-α and importin-β (BANGS et al. 1998; SHAH et al. 1998). Since this interaction only takes place in the absence of karyophilic proteins, it is conceivable that Tpr is involved in the recycling of importin-α and importin-β rather than in the nuclear import of karyophilic proteins (BANGS et al. 1998; SHAH et al. 1998). Recently, two yeast homologues of Tpr have been identified, termed Mlp1p and Mlp2p (STRAMBIO-DE-CASTILLIA et al. 1999; KOSOVA et al. 1999), and, using immunogold-EM analysis, both have been located to the intranuclear filaments, which connect the NPC with the nuclear interior (Table 1; Fig. 2). Similar to Tpr, evidently Mlp1p and Mlp2p are not involved in nuclear protein import. Tpr has also been suggested to participate in mRNA export (BANGS et al. 1998), while this transport activity has not been evaluated for Mlp1p and Mlp2p (STRAMBIO-DE CASTILLA et al. 1999).

Nup145p, a yeast nucleoporin involved in mRNA export (FABRE et al. 1994), is cleaved in vivo to yield two functionally distinct domains (DOCKENDORFF et al. 1997; EMTAGE et al. 1997; TEIXEIRA et al. 1997). After cleavage, the C-terminal domain of Nup145p, termed C-Nup145p, becomes assembled into the Nup84p complex of the yeast NPC, which locates to its cytoplasmic fibrils (Fig. 2; see also TEIXEIRA et al. 1997; S. Siniossouglou and E. Hurt, personal communication). Not only is C-Nup145p involved in mRNA export, but it is necessary for cell viability and proper NPC distribution within the NE (DOCKENDORFF et al. 1997; EMTAGE et al. 1997; TEIXEIRA et al. 1997). In contrast, the function of N-Nup145p, the N-terminal do-

Table 1. Some newly identified nucleoporins

Name[a]	Putative homologue(s)	Motifs[b]	Location	Properties and function	References
h PBC68	–	–	Nucleoplasmic face of the NPC	Colocalizes with mitotic spindle involved in primary biliary cirrhosis	THEODOROPOULOS et al. (1999)
r Nup96	Sc C-Nup145p	–	Nucleoplasmic face of the NPC	Generated by autoproteoloytcally in vivo cleavage of a Nup98-Nup96	FONTOURA et al. (1999)
				Precursor; in complex with Nup107 and two Sec13-related proteins	ROSENBLUM and BLOBEL 1999
Sc Nup92p	r, h p205	–	Nuclear basket filaments	Necessary for assembly of Nup49p, Nup57p, Nup82p, and Nic96p into the NPC	KOSOVA et al. (1999)
Sc Mlp1p/Sc Mlp2p	r, h, X, D Tpr	Coiled-coil P/F-rich region	Nuclear basket and intranuclear filaments	C-terminal of Mlp1p, responsible for nuclear localization; Mlp1p and Mlp2p involved in nuclear protein import	STRAMBIO-DE-CASTILLIA et al. (1999)
Sp Nup124p	–	FG repeats	Unknown	Reduced level of *S. pombe* Tf1 transposition in *nup124-1* strain; role in nuclear import of Tf1 Gag protein	BALASUNDARAM et al. (1999)

Nomenclature used: *h*, human; *Nup*, nuclear pore protein; *r*, rat; *Sc*, *Saccharomyces cerevisiae*; *Sp*, *Schizosaccharomyces pombe*; *X*, *Xenopus laevis*; *NES*, nuclear export sequence. *Coiled-coil*, predicted parallel, two-stranded α-helical structure made of heptad repeats;
[a] Numerical assignment reflects either the predicted molecular mass (in kDa) or the genetic identification;
[b] FXFG, GLFG, FG, GSXS, GSSX, GFXS, PA, and WD repeat motifs represented by single-letter code for amino acids.

main of Nup145p, remains to be established. Recently, the vertebrate homologue of the yeast Nup145p ~186-kDa precursor, has been identified (Table 1; FONTOURA et al. 1999). Nup96 and Nup98 are generated by cleavage of the 186-kDa precursor by an autoproteolytic process without participation of any endogenous proteases (ROSENBLUM and BLOBEL 1999). Nup98 appears to be the homologue of N-Nup145p, whereas Nup96 appears to be the homologue of C-Nup145p. However, the location of these two functional homologues remains controversial: whereas C-Nup145p is located at the cytoplasmic fibrils of the yeast NPC (Fig. 2; S. Siniossouglou and E. Hurt, personal communication), Nup98 and Nup96 have both been located to the nuclear basket (Fig. 2; see RADU et al. 1995; FONTOURA et al. 1999). At least for Nup96, this controversy might be in part due to the cross-reactivity of the antibody used for its immunolocalization, which cross-reacts with the nucleoplasm. C-Nup145p, on the other hand, has been localized via a protein-A fusion protein with an antibody directed against the protein-A tag, a method that evidently does not cause any significant mislocalization of the resulting fusion protein (FAHRENKROG et al. 2000a). Therefore, the definite locations of vertebrate Nup96 and Nup98 and their yeast homologues C-Nup145p and N-Nup145p remain to be established.

Vertebrate Nup98 is a nucleoporin that appears to be primarily involved in distinct RNA export pathways, i.e., the export of snRNAs, 5S RNA, rRNA and mRNA, but also in nuclear growth and replication (POWERS et al. 1995, 1997; RADU et al. 1995). A number of recent publications have documented that the FG repeats of Nup98 are involved in chromosomal translocations causing acute myeloid leukemia (AML; NAKAMURA et al. 1996; ARAI et al. 1997; RAZA-EGLIMEZ et al. 1998; BORROW et al. 1999; IKEDA et al. 1999; KASPER et al. 1999; KWONG and PANG 1999; NAKAMURA et al. 1999; NISHIYAMA et al. 1999) and also acute lymphocytic leukemia (ALL; HUSSEY et al. 1999). In this context, the resulting chimeric proteins consisting of the FG repeat domains of Nup98 fused to distinct homeobox proteins, e.g., HOXA9 and HOXD13, might act as oncogenic transcription factors (KASPER et al. 1999; NAKAMURA et al. 1999), whereas *NUP98* acts as a potential tumor suppressor gene (BORROW et al. 1999; NAKAMURA et al. 1999).

Evidently, Nup98 is not the only nucleoporin involved in pathogenic processes. For example, Nup155, a vertebrate nucleoporin of unknown function, might have a possible role in mental and developmental retardation, as suggested from the genomic location of the human *NUP155* gene on chromosome band 5p13 (ZHANG et al. 1999a). Moreover, the integral membrane protein gp210 is implicated in the autoimmune disease primary biliary cirrhosis (PBC; reviewed in COURVALIN and WORMAN 1997). Accordingly, 25% of PBC patients produce auto-antibodies directed against gp210, predominatly against a distinct 15-amino acid segment residing within the cytoplasmic, C-terminal domain of gp210 (NICKOWITZ and WORMAN 1993; reviewed in COURVALIN and WORMAN 1997). Recent data have provided evidence that gp210 is not the only nucleoporin involved in PBC. Some patients with PBC also exhibit auto-antibodies against p62, one of the first-identified and best-characterized vertebrate nucleoporins (reviewed in PANTÉ and AEBI 1996a; STOFFLER et al. 1999a; KINOSHITA et al. 1999), as well as against the newly identified human nucleoporin PBC68 (THEODOROPOULOS et al. 1999).

3 The Nuclear Pore Complex Is a Dynamic Structure

Immunogold-EM revealed that many nucleoporins exhibit multiple locations within the NPC (GUAN et al. 1995; GRANDI et al. 1997; FAHRENKROG et al. 1998, 2000a; STRAHM et al. 1999). These multiple locations indicate that the NPC might be a dynamic rather than a static structure. In this context, Nsp1p and Nic96p are both located about the cytoplasmic and nuclear periphery of the central channel of the NPC and near or at the distal ring of the nuclear basket (Fig. 2; see also FAHRENKROG et al. 1998, 2000a). Based on these distinct locations, they might function as docking or parking sites at the cytoplasmic or nuclear "portal" to the

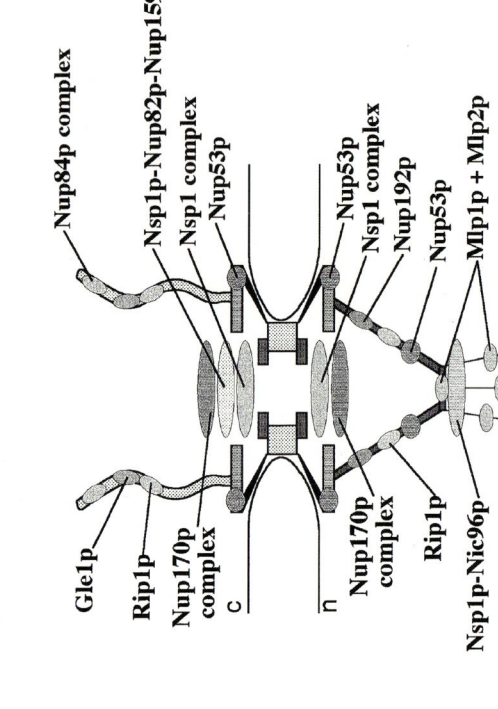

central channel so as to concentrate and/or "funnel" the cargo complex before it traverses the NPC in either direction. Furthermore, at their nuclear locations Nsp1p and Nic96p might interact with the cargo complex after it exits the central channel so as to mediate its dissociation: (1) for release of the cargo into the nucleus, and (2) for recycling the transport factor(s) back into the cytoplasm. Alternatively, Nsp1p and Nic96p might actually represent mobile nucleoporins that shuttle across the NPC from its cytoplasmic to its nuclear periphery by associating with the cargo complex, i.e., via soluble transport factors that mediate nuclear protein import. Such a direct interaction of Nsp1 with distinct import factors, for example, importin-β or the small GTPase Ran (STOCHAJ et al. 1998; SEEDORF et al. 1999), has indeed been demonstrated, suggesting that Nsp1p may accompany the cargo complex for some distance along its path through the NPC. However, no direct interaction of Nic96p with any transport factors has yet been demonstrated.

Shuttling of nucleoporins has recently been demonstrated for three vertebrate nucleoporins, Nup98, Nup153p and CAN/Nup214 (BOER et al. 1997; NAKIELNY et al. 1999; ZOLOTUKHIN and FELBER 1999). Nup98, which resides at the nuclear basket and therefore inside the nucleus (see above), translocates from the nucleus to the cytoplasm of HeLa cells upon addition of the transcription inhibitor actinomycin D (ZOLOTUKHIN and FELBER 1999). Nup153 has been located to the distal ring of the nuclear basket in the vertebrate NPC (PANTÉ et al. 1994) and is involved

◀──────────────────────────────────────

Fig. 2a,b. A summary of the immunolocalization of nucleoporin epitopes within the 3-D architecture of the vertebrate (*left*) and the yeast (*right*) nuclear pore complex (NPC). In vertebrates CAN/Nup214 and RanBP2 exhibit epitopes at the cytoplasmic fibrils (reviewed in PANTÉ and AEBI 1996; STOFFLER et al. 1999a). The p62 complex, consisting of p62, p58, p54, and p45, exhibits epitopes at the cytoplasmic and the nuclear periphery of the central gated channel (reviewed in PANTÉ and AEBI 1996; STOFFLER et al. 1999a); additionally, p62 exhibits an epitope at the nuclear basket (see FAHRENKROG et al. 1998; STOFFLER et al. 1999a). Nup93 epitopes are located at the nuclear periphery of the central channel and at the nuclear basket (GRANDI et al. 1997), Nup153, Nup98, Nup96 and Tpr exhibit epitopes at the nuclear basket, the latter two also at the intranuclear filaments (reviewed in PANTÉ and AEBI 1996; STOFFLER et al. 1999a; ZIMOWSKA et al. 1997; BANGS et al. 1998; CORDES et al. 1998; FONTOURA et al. 1999). The transmembrane proteins gp210 and POM121 are predicted to have epitopes in the lumen of the nuclear envelope (NE) and within the NPC proper, respectively (reviewed in PANTÉ and AEBI 1996; STOFFLER et al. 1999a). In yeast, the nucleoporins of the Nup84p complex, i.e., C-Nup145p, Nup120p, Nup85p, Nup84p, Sec13p, and Seh1p, display epitopes at the cytoplasmic fibrils (S. Siniossoglou and E. Hurt, personal communication). Gle1p and Rip1p epitopes reside at the cytoplasmic fibrils, Rip1p additionally at the nuclear basket (STRAHM et al. 1999). The epitopes of the Nsp1p complex (Nsp1p-Nup49p-Nup57p-Nic96p) are located at the cytoplasmic and the nuclear periphery of the central channel (FAHRENKROG et al. 1998). Moreover, Nsp1p and Nic96p exhibit epitopes at the distal ring of the nuclear basket (FAHRENKROG et al. 1998). Epitopes of the Nsp1p-Nup82p-Nup159p complex are displayed at the cytoplasmic periphery of the central channel (BELGAREH et al. 1998; FAHRENKROG et al. 1998; HURWITZ et al. 1998). The Nup170p complex, i.e., Nup188p, Nup170p, Nup157p, Nup59p, and Nup53p, exhibit epitopes at the cytoplasmic and nuclear face of the NPC core (reviewed in PANTÉ and AEBI 1996; STOFFLER et al. 1999a; MARELLI et al. 1998; FAHRENKROG et al. 2000b). Whereas the location of Nup188p, Nup170p, Nup157p, and Nup59p has not been assigned to NPC substructures, Nup53p is located at the cytoplasmic and the nuclear face of the central framework and at the fibrils of the nuclear basket (FAHRENKROG et al. 2000b). The yeast homologues of Tpr, named Mlp1p and Mlp2p reside at the nuclear basket and the intranuclear filaments (STRAMBIO-DE-CASTILLIA et al. 1999; KOSOVA et al. 2000). However, since all these localization studies have been performed with antibodies directed against epitopes of nucleoporins or tags fused to nucleoporins, we are only at the beginning of understanding the complete 3-D molecular architecture of the NPC. For a more complete immunolocalization of yeast nucleoporin epitopes see ROUT et al. (2000)

in mRNA export and protein import (BASTOS et al. 1996; SHAH et al. 1998). Evidently, Nup153, upon association with the export cargo complex, shuttles from the nuclear periphery of the NPC to its cytoplasmic face (NAKIELNY et al. 1999). CAN/Nup214, on the other hand, is a constituent of the cytoplasmic fibrils of the vertebrate NPC (PANTÉ et al. 1994) and has been demonstrated to be involved in protein import and mRNA export (VAN DEURESEN et al. 1996; FORNEROD et al. 1997). It has been documented that CAN/Nup214 changes its location from the cytoplasmic fibrils to the nuclear basket upon overexpression in HeLa cells, suggesting that the location of a given nucleoporin may also depend on its expression level within the cell (BOER et al. 1997).

Nup49p and Nup57p both reside about the cytoplasmic and the nuclear periphery of the central channel and evidently represent more stationary nucleoporins (see Fig. 2; see also FAHRENKROG et al. 1998). Their location suggests that both participate in the mid-steps of protein import and/or RNA export, thereby acting as docking or parking sites for the cargo complex immediately before and after translocation through the central channel.

Green fluorescent protein (GFP)-tagged nucleoporins combined with mating assays in yeast have demonstrated that not only individual nucleoporins, but even fully assembled NPCs may move within the NE (BELGAREH and DOYE 1997; BUCCI and WENTE 1997). Moving of NPCs within the NE is a feature that is shared with the spindle pole body (SPB). Both particles also share at least some proteins: Ndc1p in *Saccharomyces cerevisiae* and Cut11p in *Schizosaccharomyces pombe* locate within both the NPC and the SPB, as determined by immunofluorescence microscopy (CHIAL et al. 1998; WEST et al. 1998). PBC68 (Table 1; see above) associates with the mitotic spindle, suggesting that nucleoporins and probably proteins of the NE too are actively sorted during mitosis by transient anchoring to spindle microtubules (THEODOROPOULOS et al. 1999). A connection between the NPC and the actin cytoskeleton has also been demonstrated by the fact that mutations of the yeast divergent actin gene *ACT2* caused defects in NPC structure and nuclear protein import (YAN et al. 1997). Recently, a link between sporulation and NE/NPC organization was documented (SINIOSSOGLOU et al. 1999): the two integral membrane proteins Spo7p and Nem1p, as well as the nucleoporins Nup84p, Nup85p and Nup133p, are evidently essential for sporulation. By synthetic lethality, *NUP84* interacts genetically with *SPO7* and *NEM1*, and most likely also physically, albeit transiently (SINIOSSOGLOU et al. 1999). Moreover, disruption of *SPO7* and *NEM1* caused drastic alternations of the nuclear morphology. Taken together, NPC assembly appears to be coupled to sporulation and to the biogenesis of other cell organelles and compartments, such as the SPB and the actin cytoskeleton.

4 Interactions Between Nucleoporins and Viral Proteins

For replication, many viruses depend on nuclear host cell factors, so that the viral genome has to enter the nucleus (reviewed in KASAMATSU and NAKANISHI 1998;

WHITTAKER and HELENIUS 1998; IZAURRALDE et al. 1999). Hence in non-dividing cells viruses have to traverse the NE via the NPC. As a consequence, much effort has been spent to identify and characterize signals and factors that mediate the import and the export of viral proteins, as well as the nucleoporins that are involved in the corresponding transport pathways. How viral proteins interact with nucleoporins while traversing the NPC is best understood for the import of the HIV-1 protein Vpr and the export of the HIV-1 Rev protein, although some controversies remain (see FOUCHIER et al. 1998; JENKINS et al. 1998; POPOV et al. 1998; VODICKA et al. 1999).

Vpr appears to interact with the FG repeats of several vertebrate and yeast nucleoporins, including POM121, p54, Nsp1p and Nup1p (JENKINS et al. 1998; POPOV et al. 1998). p54 and Nsp1p are located about the cytoplasmic and nuclear periphery of the central channel, and Nsp1p additionally near or at the distal ring of the nuclear basket (FAHRENKROG et al. 1998; HU and GERACE 1998; FAHRENKROG et al. 2000a; reviewed in STOFFLER et al. 1999a), suggesting that both nucleoporins act as docking or parking sites for Vpr before and after traversing the central channel. More specifically, Nsp1p might be involved in the release of Vpr from the NPC en route to the nucleus. Moreover, Vpr also clearly interacts with importin-α, whereas its interaction with importin-β remains controversial (see FOUCHIER et al. 1998; JENKINS et al. 1998; POPOV et al. 1998; VODICKA et al. 1999). In this context, it has been speculated that Vpr might act as a specialized importin-β transport factor (JENKINS et al. 1998; VODICKA et al. 1999) so as to overcome the import pathway usually used by karyophilic proteins, a hypothesis that is supported by the finding that the NLS of Vpr is distinct from the classical NLS and the M9-like NLS (FOUCHIER et al. 1998; JENKINS et al. 1998). As an alternative, other authors have suggested that Vpr function depends on importin-β (POPOV et al. 1998).

The HIV-1 Rev protein harbors a nuclear export signal (NES) whose determination led to the identification of the first nuclear export factor, termed CRM1 (exportin; in yeast called Crm1p or Xpo1p; reviewed in IZAURRALDE and ADAM 1998; STUTZ and ROSBASH 1998). Similar to Vpr, Rev interacts with a number of nucleoporins, namely, CAN/Nup214 and its yeast homologue Nup159p, Rip1p in vertebrates and yeast (also termed NLP1, hCG1 or Nup42p; BOGERD et al. 1998; FARJOT et al. 1999; FLOER and BLOBEL 1999; STRAHM et al. 1999; ZOLOTUKHIN and FELBER 1999), and with the vertebrate Nup98 (ZOLOTUKHIN and FELBER 1999). In agreement with these interactions of Rev, a truncated CAN/Nup214 leads to the inhibition of Rev-dependent late gene expression of HIV-1 (BOGERD et al. 1998).

Vertebrate CAN/Nup214 and its yeast homologue Nup159p are located at the cytoplasmic fibrils of the vertebrate NPC and the cytoplasmic face of the yeast NPC, respectively (see Fig. 2; see also PANTÉ et al. 1994; KRAEMER et al. 1995), suggesting that CAN/Nup214 and Nup159p are involved in a late step of Rev export, i.e., its release from the NPC into the cytoplasm. The interaction between Rev and Nup98 suggests that Nup98, a component of the nuclear basket (see Fig. 2; see also RADU et al. 1995), participates in the docking of Rev to the NPC. Rip1p is located at the cytoplasmic fibrils and the nuclear basket of the yeast NPC (STRAHM et al. 1999), indicating that Rip1p might be involved in distinct steps of

Rev export, i.e., the docking of Rev to the NPC at the nuclear basket, as well as its release from the cytoplasmic fibrils into the cytoplasm.

The influenza virus NEP protein (formerly termed NS2) harbors a Rev-like NES and strongly interacts with Rip1p and its human homologue, suggesting that Rip1p functions as a docking and release site for NEP at the NPC (O'NEILL et al. 1998). Another viral protein interacting with CAN/Nup214 is SM, an Epstein-Barr virus nuclear protein. Its translocation to the cytoplasm depends on the presence of CRM1 and the interaction with CAN/Nup214 at the cytoplasmic fibrils, which might be the termination site of SM protein export (BOYLE et al. 1999). CAN/Nup214 is also involved in the export of unspliced transcripts of the type D retroviruses, which contain a sequence termed constitutive transport element (CTE). The nuclear export of CTE-containing RNAs is mediated by a nuclear export factor called TAP (GRÜTER et al. 1998; reviewed in STRAMBIO-DE-CASTILLIA and ROUT 1999). TAP does not belong to the importin-β family of transport factors and uses an export route that is different from the CRM1 pathway (BOGERD et al. 1998; OTERO et al. 1998) since the NES of TAP is different from the Rev-like NES (BEAR et al. 1999; KANG and CULLEN 1999). TAP interacts at the cytoplasmic fibrils with the human nucleoporins CAN/Nup14 and hCG1 (KATAHIRA et al. 1999), suggesting that this interaction might occur as a late step of TAP-mediated CTE-dependent export before release of the RNAs into to the cytoplasm.

5 Nuclear Pore Complex Architecture

Determining the 3-D and molecular architecture of the NPC is a prerequisite for a structure-based functional analysis of nucleo-cytoplasmic transport at the molecular level. The 3-D reconstructions using amphibian NEs have provided considerable insights into the 3-D architecture of the NPC, and over the past several years there has been significant progress in identifying and characterizing distinct NPC subcomplexes (reviewed in STOFFLER et al. 1999a).

The recently solved 3-D structure of isolated yeast NPCs by electron cryo-microscopy revealed a surprisingly flat, $\sim$822 symmetric spoke complex embracing a central plug, with no clear indication of a cytoplasmic or a nuclear ring being attached to it (YANG et al. 1998). The mass of the yeast NPC used for the 3-D reconstruction was determined to be $\sim$60MDa, a value consistent with its smaller linear dimensions (see below). However, by comparing the overall dimensions of the yeast and vertebrate 3-D reconstructions, the yeast NPC must be significantly more compact than the vertebrate NPC, otherwise its mass would only amount to $\sim$30MDa. While this marks a promising start, it is only the first word on the 3-D architecture of the yeast NPC.

By comparing yeast and vertebrate NPC structure, it was noted that the overall 3-D architecture of the NPC is evolutionarily conserved from yeast to higher eukaryotes. Whereas the yeast NPC's linear dimensions were determined to be about 15% smaller than those of *Xenopus* oocyte NPCs, the yeast NPC also revealed cytoplasmic fibrils and a nuclear basket (FAHRENKROG et al. 1998). Moreover, in the context of investigating the dynamic behavior of NPC structure during Balbiani ring particle translocation in salivary gland cells of the insect *Chironomus thummi* (KISELEVA et al. 1998), it was found that in most respects *Chironomus* NPCs were similar to amphibian NPCs, thus suggesting a strong evolutionary conservation of NPC architecture between invertebrates and vertebrates.

6 Functional States of the Nuclear Pore Complex

Since the NPC appears to be a rather dynamic structure (see above; reviewed in STOFFLER et al. 1999a), identification and characterization of distinct functional states of the NPC will be a prerequisite to pursue a structure-based functional analysis of nucleocytoplasmic transport. For example, significant structural changes of the NPC can be induced by cellular signals such as calcium or nucleotides (JARNIK and AEBI 1991; Fig. 3c,d and Fig. 4b,c). Remarkably, EM has indicated that NTF2/p10, a nuclear import factor, might regulate nucleocytoplasmic transport by modulating the functional size of the gated channel within the NPC during oogenesis (FELDHERR et al. 1998). Yet, it is unlikely that EM will be the tool of choice to dissect the more dynamic features of macromolecular machines such as the NPC and to directly correlate structural with functional states in situ. Atomic force microscopy (AFM), on the other hand, has now evolved to the point where it is possible to record the surface topography of native biological matter in its physiological buffer environment at molecular detail and to directly correlate structural changes with distinct functional states (Fig. 3; see also STOFFLER et al. 1999b; reviewed by ENGEL et al. 1999).

Along these lines, the effect of ATP on nuclear pore conformation in isolated NEs from *Xenopus laevis* oocytes was investigated using AFM and revealed overall size and shape changes of NPCs following addition of ATP (RAKOWSKA et al. 1998). AFM experiments to examine the effect of calcium on the structure of cardiac NPCs were also performed (PEREZ-TERZIC et al. 1999). However, in none of these examples were fully native NPCs explored by AFM: the material was chemically fixed, exposed to detergents, or dehydrated and rehydrated at some stage during its preparation to improve the adhesion of the sample to the specimen support. Hence, it is unclear whether the distinct morphologies depicted in these studies do indeed represent bona fide structural changes or functional states in response to adding effectors or ligands, or whether they are merely due to specimen preparation effects, for example, involving differential extraction of labile or weakly bound NPC constituents, or deformations caused by surface tension.

In contrast, time-lapse AFM images of the nuclear face of native NEs (i.e., without detergent treatment, chemical fixation, or dehydration/rehydration during preparation steps) kept functional in physiological buffer (Fig. 3a,b) has depicted

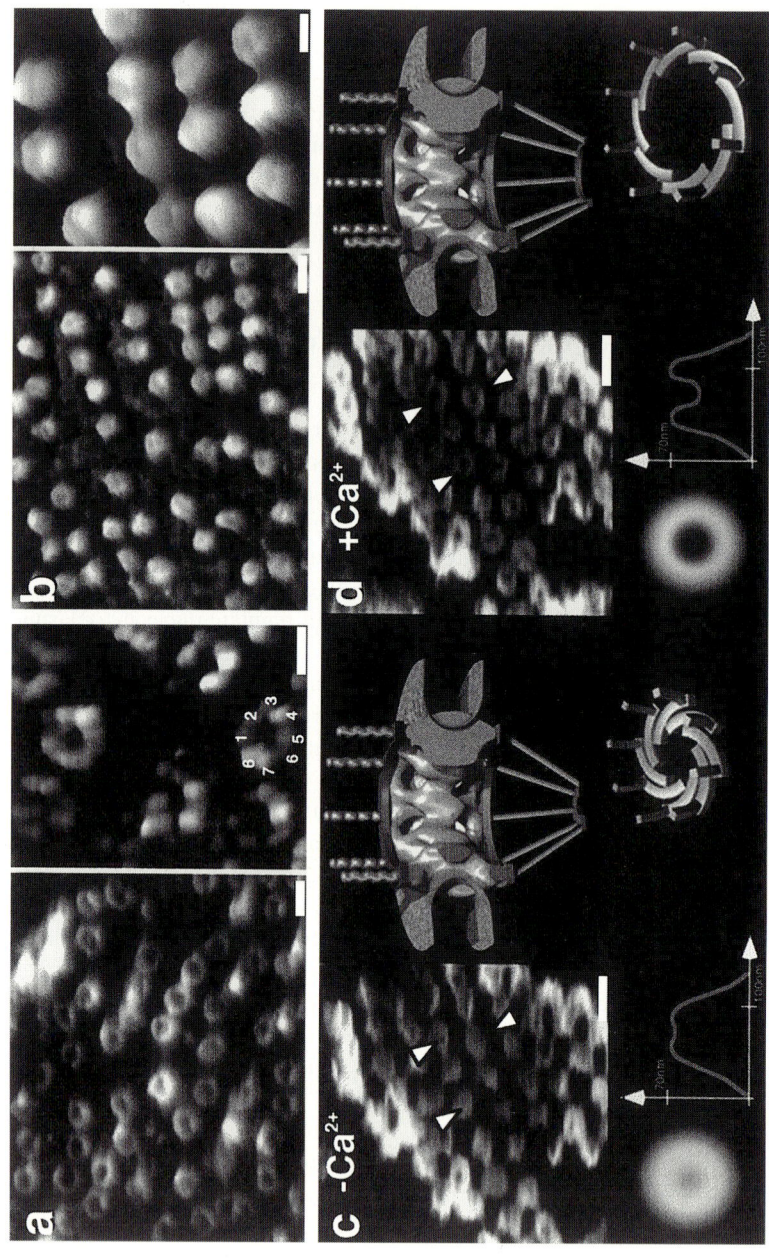

Fig. 3a–d. Visualization of native (i.e., completely unfixed, without de- or rehydration and detergent treatment) NPCs by atomic force microscopy (AFM). The NPCs were kept functional in near-physiological buffer medium (STOFFLER et al. 1999b). Corresponding AFM images revealed a distinct morphology for the cytoplasmic (**a**) and the nuclear face (**b**) of *Xenopus* oocyte NEs. The *right inset* in (**a**) depicts a high-magnification view of the cytoplasmic face so that the 8-fold rotational symmetry of individual NPCs is resolved. **b** On the nuclear face, note the "remnants" of the nuclear lamina depicted in areas devoid of NPCs. The *right inset* in (**b**) reveals a high-magnification view of the nuclear face. **c** Visualization of the reversible calcium-mediated opening (i.e., $+100\mu m$ Ca^{2+}) and closing (i.e., $-Ca^{2+}$ and $+1$ mM EGTA) of the nuclear baskets (i.e., via dilatation of their distal ring, which may act as an iris-like aperture) by time-lapse AFM. The same specimen area has been imaged in two distinct conformational states, with three corresponding NPCs being marked by *white arrowheads*. For a more quantitative comparison of the "closed" and "open" states, 30 corresponding NPCs have been aligned and averaged, and their average radial height profiles computed. The 3-D models depict the distal ring which may act as an iris-like aperture, as proposed by PANTÉ and AEBI (1996a), opening upon addition of micromolar amounts of calcium and closing upon removal of calcium. *Scale bars* **a**, **b** 100nm, **c**, **d** 200nm

the repeated opening and closing of the nuclear basket of individual NPCs at its distal end in response to adding or removing calcium (Fig. 3c,d; see also STOFFLER et al. 1999b). The observed structural changes were such that a 20–30-nm diameter opening at the distal end of the nuclear basket occurred upon addition of calcium without, however, affecting the overall height and shape of the basket (Fig. 3d). This reversible structural change of the nuclear basket may be interpreted in terms of its distal ring acting as an "iris-like" diaphragm, as suggested previously by PANTÉ and AEBI (1996a), which is closed in the absence of calcium and opens upon addition of calcium (Fig. 3c,d). Hence the question arises as to whether the distal ring might be directly involved in mediating nucleocytoplasmic transport. In support of this possibility, Nup153, a nucleoporin that is a constituent of the nuclear basket and which displays an epitope near or at the distal ring (PANTÉ et al. 1994), has been documented to play an important role in protein import as a terminal docking site (SHAH et al. 1998), as well as an initial docking site in mRNA export (BASTOS et al. 1996). Moreover, it is conceivable that Nup153, which forms an octameric complex (PANTÉ et al. 1994), acts as a structural scaffold of the eight-fold symmetric distal ring (PANTÉ et al. 1996b). Similar to the cytoplasmic fibrils, the distal ring might act as a docking or even a gating site for cargo transported in and out of the nucleus. In contrast to the calcium-induced effects observed on the nuclear face of the NE, the cytoplasmic face appeared rather unaffected by the same calcium regimen (STOFFLER et al. 1999b).

The AFM, however, is a "surface sensor" and thus allows recording of the surface topography of a given sample rather than its internal structure. As the structural change monitored on the nuclear face of the NPC in response to adding or removing calcium clearly occurred at its highest elevation above the nuclear membrane surface, it had to involve the distal ring of the nuclear basket. Obviously, based on the AFM, one cannot exclude the possibility that changes also occur in the interior of the NPC (JARNIK and AEBI 1991; STOFFLER et al. 1999b; WANG et al. 1999) that cannot be detected by AFM since they are inaccessible to the scanning tip. To explore the possibility of internal conformational changes of the NPC in

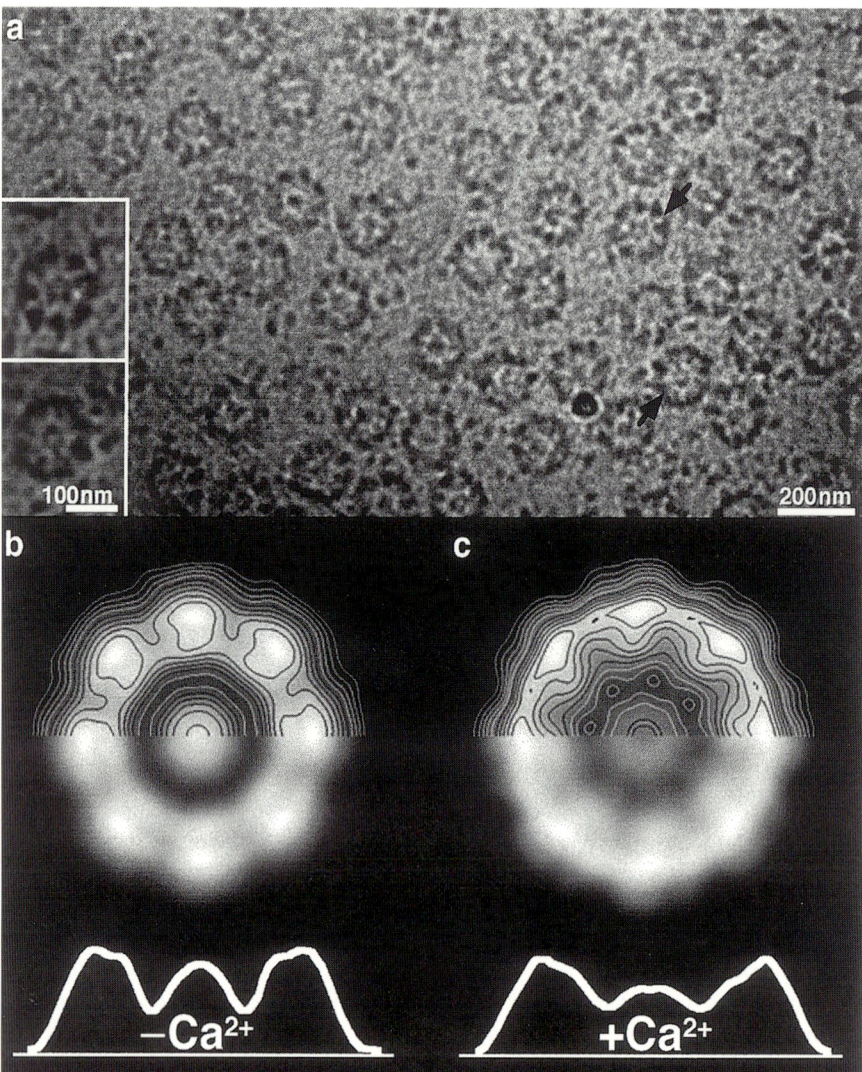

Fig. 4. a Energy-filtering transmission electron microscope (EFTEM) images of native, unstained and unfixed nuclear envelopes (NEs) embedded in a thick, amorphous ice layer (i.e., ~200nm). The *inset* in (**a**) depicts higher magnification views of two NPCs, revealing details of their 8-fold rotational symmetric central framework as well as NPC-associated filaments such as the nuclear basket with its terminal ring. **b, c** Single-particle averages, both displayed as gray-level ±contour representations and radial mass density profiles of 100 NPCs each in the absence (**b**) and presence (**c**) of calcium, revealed significant structural rearrangements within the entire NPC

response to effectors such as calcium or ATP, cryo-transmission electron microscopy (cryo-TEM) of completely unfixed/unstained, spread NEs embedded in a thick layer (i.e., ~250nm) of amorphous ice was performed (Fig. 4; see also

STOFFLER et al. 1999b; Stoffler et al., manuscript in preparation). Averages of NPC projection images, each of a sample incubated with and without a given effector prior to freezing, revealed subtle but significant structural rearrangements within the entire NPC (Fig. 4b,c). Besides changes involving the distal ring of the nuclear basket, other changes may involve the peripheral channels of the NPC central framework (Fig. 4b,c). Whereas a detailed understanding of the structural changes occurring in response to adding or removing effectors to the NPC has to await tomographic reconstructions (Stoffler et al., manuscript in preparation), the cryo-TEM-based projection data are clearly consistent with the interpretation of the corresponding AFM data, for example, that the opening and closing of the nuclear basket's distal ring resembles an iris-like diaphragm (Fig. 4b,c; see also STOFFLER et al. 1999b).

7 Controversial Nuclear Pore Complex Components: the "Central Plug"

Three-dimensional reconstructions of detergent-extracted *Xenopus* NPCs (but not of membrane-bound NPCs) revealed a massive barrel-shaped particle occupying the central pore along its entire length (AKEY and RADERMACHER 1993). A similar "central transporter" was also revealed in the 3-D reconstruction of yeast NPCs (YANG et al. 1998). A model of the transporter substructure was proposed based on transmission EM and field emission in-lens scanning EM (FEISEM) data of *Chironomus* NPCs. In this model, the transporter consists of two central cylinders and two globular assemblies that undergo conformational variations during Balbiani ring particle translocation (KISELEVA et al. 1998). It was presumed that this structure might actually represent the transport machinery of the NPC. However, it should be kept in mind that, depending on the isolation and specimen preparation procedures employed, both the abundance and appearance of this transporter are highly variable (reviewed by PANTÉ and AEBI 1996a). Furthermore, other NPC components, such as the distal ring of the nuclear basket, which might have been squeezed into the central pore during specimen preparation may contribute significantly to what in 3-D reconstructions of detergent-extracted or otherwise distorted NPCs appears as the central plug. Moreover, the multiple locations of some nucleoporins (for example, p62) indicate that these may represent mobile – rather than stationary – NPC components (see above) accompanying the cargo complex for some distance while traversing the NPC (reviewed by STOFFLER et al. 1999a). Hence, it is conceivable that the central plug might represent a mobile NPC moiety which, at least in part, consists of cargo caught in transit. Therefore it remains elusive to what extent the central plug or transporter is actually a stationary moiety of the NPC and/or to what extent it represents cargo associated with transport factors and/or "mobile" nucleoporins.

8 Nuclear Pore Complex Disassembly and Reassembly

A better understanding of the interactions between different nucleoporins may be gained by investigating the pathways and molecular mechanisms involved in NPC disassembly and reassembly. For example, during mitosis in higher cells the double-membraned NE and the NPCs disassemble and are dispersed throughout the mitotic cytoplasm (MACAULAY 1995; FAVREAU et al. 1996; COLLAS 1998).

It was suggested that NE disassembly correlates with phosphorylation of proteins of the inner nuclear membrane, the nuclear lamins, and several nucleoporins including CAN/Nup214, Nup98, Nup358, Nup153 and gp210 (reviewed by DOYE and HURT 1997). Recently, it was demonstrated that NPCs may play a structural and functional role during NE disassembly in that phosphorylation of nucleoporins in late prophase may alter pore-membrane and pore-lamina interactions (COLLAS 1998). Moreover, it was shown that the absence of NPCs or inhibition of pore function prevented NE and lamina disassembly (COLLAS 1998). Hence, it was suggested that NPCs may not merely provide access of soluble components implicated in NE disassembly, but perhaps concentrate these factors to a certain threshold at the level of the NE or inside the nucleus (COLLAS 1998).

Whereas the amphibian NPC is disassembled into several soluble subcomplexes during mitosis (MACAULAY 1995), the biochemical behavior of different NPC components of mammalian somatic cells during mitosis remains unclear. Recent evidence suggests that mammalian somatic NPCs disassemble into at least three subcomplexes during mitosis (MATSUOKA et al. 1999). Whereas subcomplex A contains Nup214/p62 and subcomplex B also contains p62, subcomplex C consists of the nucleoporins Nup98 and RAE1 (i.e., the human homologue of yeast Rae1p/Gle2p).

The steps of NPC assembly following mitosis and the nature of NPC structural intermediates are still poorly understood. However, recent observations defined a stepwise postmitotic association of nucleoporins with chromosomes (BODOOR et al. 1999). Accordingly, the process starts in late anaphase with the recruitment of Nup153 and continues until late telophase. The order in which the various nucleoporins reappear at the reforming NE is Nup153, POM121, p62 and CAN/Nup214, followed by gp210 and Tpr (BODOOR et al. 1999). Furthermore, these findings support a model in which POM121 potentially defines sites of NPC assembly, which subsequently recruit additional nucleoporins in a hierarchic order (BODOOR et al. 1999).

Apoptosis represents a controlled cell death mechanism that is required to maintain tissue homeostasis. It can be triggered by exposing cells to a variety of drugs (reviewed by WERTZ and HANLEY 1996) or by virus infection (reviewed by ROULSTON et al. 1999). The apoptotic breakdown of the NE was investigated under conditions in which a caspase-3-like protease seemed to play a major role (BUENDIA et al. 1999). Using different apoptotic inducers and different cell lines, it was demonstrated that lamin B, LAP2 (an inner nuclear membrane protein) and Nup153 were selectively and conservatively proteolyzed (i.e., yielding fragments of a specific size; BUENDIA et al. 1999). Based on these findings, it was suggested that their cleavage might allow both detachment of the NE from the chromatin and clustering of the NPCs in the plane of the membrane, two morphological hallmarks of apoptosis observed in that study.

9 Conclusions

Despite much recent progress made toward a better understanding of the 3-D architecture of the NPC, the identity, molecular composition and functional significance of some of its components have remained controversial, for example, the central plug. Nevertheless, mapping nucleoporins within the 3-D architecture of both the vertebrate and the yeast NPC and depicting distinct interactions between nucleoporins and transport factors or viral proteins have increased significantly our understanding on the role of particular nucleoporins in nucleocytoplasmic transport. However, the key to a more complete understanding as to how transport substrates such as viral proteins move across the NPC will only be provided by solving the atomic structure of nucleoporins. This has recently been achieved by X-ray crystallography for several transport factor complexes, some with a bound nuclear localization signal (SCHEFFZEK et al. 1995; CONTI et al. 1998; STEWART et al. 1998; CHOOK and BLOBEL 1999; VETTER et al. 1999a). Yet, except for the Ran-binding domain of RanBP2 (i.e., in complex with Ran bound to the non-hydrolysable GTP analogue GppNHp; VETTER et al. 1999b), nucleoporins are still awaiting atomic structure determination.

Acknowledgements. We would like to thank Robert Wyss for preparing the NPC schemes used in Figs. 1 and 2. This work was supported by research grants from the Swiss National Science Foundation (SNF) and the Human Frontier Science Program (HFSP), and by the Canton Basel-Stadt and the M.E. Müller Foundation of Switzerland.

References

Adam SA (1999) Transport pathways of macromolecules between the nucleus and the cytoplasm. Curr Opin Cell Biol 11:402–406

Akey CW, Radermacher M (1993) Architecture of the *Xenopus* nuclear pore complex revealed by 3-dimensional cryo-electron microscopy. J Cell Biol 122:1–19

Arai Y, Hosoda F, Kobayashi H, Arai K, Hayashi Y, Nanao K, Kaneko Y, Ohki M (1997) The inv(11)(p15q22) chromosome translocation of *de novo* and therapy-related myeloid malignancies results in fusion of the nucleoporin gene NUP98, with the putative RNA helicase gene, DDX10. Blood 89:3936–3944

Bailer SM, Siniossoglou S, Podtelejnikov A, Hellwig A, Mann M, Hurt E (1998) Nup116p and Nup100p are interchangeable through a conserved motif which constitutes a docking site for the mRNA transport factor Gle2p. EMBO J 17:1107–1119

Balasundaram D, Benedik MJ, Morphew M, Dang VD, Levin HL (1999) Nup124p is a nuclear factor of Schizosaccharomyces pombe that is important for nuclear import and activity of retrotransposon Tf1. Mol Cell Biol 19:5768–5784

Bangs P, Burke B, Powers C, Craig R, Purohit A, Doxsey S (1998) Functional analysis of Tpr: identification of nuclear pore complex association and nuclear localization domain and a role in mRNA export. J Cell Biol 143:1801–1812

Bastos R, Lin A, Enarson M, Burke B (1996) Targeting and function of nuclear pore complex protein Nup153. J Cell Biol 134:1141–1156

Bastos R, de Pouplana LR, Enarson M, Bodoor K, Burke B (1997) Nup84, a novel nucleoporin that is associated with CAN/Nup214 on the cytoplasmic face of the nuclear pore complex. J Cell Biol 137:989–1000

Bear J, Wei T, Zolotukhin AS, Tabernero C, Hudson EA, Felber BK (1999) Identification of novel import and export signals of human TAP, the protein that binds to the constitutive transport element of the type D retrovirus mRNAs. Mol Cell Biol 19:6309–6317

Belgareh N, Doye V (1997) Dynamics of the nuclear pore distribution in nucleoporin mutant yeast cells. J Cell Biol 136:747–759

Belgareh N, Snay-Hodge C, Pasteau F, Dagher S, Cole CN, Doye V (1998) Functional characterization of a Nup159p-containing nuclear pore subcomplex. Mol Biol Cell 9:3475–3492

Boer JM, van Deursen JMA, Huib HC, Fransen JAM, Grosveld GC (1997) The nucleoporin CAN/Nup214 binds to both the cytoplasmic and the nucleoplasmic sides of the nuclear pore complex in overexpressing cells. Exp Cell Res 232:182–185

Bodoor K, Shaikh S, Salina D, Raharjo WH, Bastos R, Lohka M, Burke B (1999) Sequential recruitment of NPC proteins to the nuclear periphery at the end of mitosis. J Cell Sci 112:2253–2264

Bogerd HP, Fridell RA, Benson RE, Hua J, Cullen BR (1996) Protein sequence requirements for function of the human T-cell leukemia virus type I Rex nuclear export signal delineated by a novel in vivo randomization-selection assay. Mol Cell Biol 16:4207–4214

Bogerd HP, Echarri A, Ross TM, Cullen BR (1998) Inhibition of immunodeficiency virus Rev and human T-cell leukemia virus Rex, but not mason-pfizer monkey virus constitutive transport element activity, by a mutant human nucleoporin targeted to Crm1. J Virol 72:8627–8635

Borrow J, Shearman AM, Stanton Jr VP, Becher R, Collins T, Williams AJ, Dubé I, Katz F, Morris C, Ohyashiki K, Toyama, K, Rowley J, Housman DE (1996) The t(7;11)(p15;p15) translocation in acute myeloid leukemia fuses the genes for nucleoporin *NUP98* and class I homeoprotein *HOXA9*. Nature Genet 12:159–167

Boyle SM, Ruvolo V, Gupta AK, Swaminathan S (1999) Association with the cellular export receptor CRM1 mediates function and intracellular localization of Epstein-Barr virus SM protein, a regulator gene expression. J Virol 73:6872–6881

Braun IC, Rohrbach E, Schmitt C, Izaurralde E (1999) TAP binds to the constitutive transport element (CTE) through a novel RNA-binding motif that is sufficient to promote CTE-dependent RNA export from the nucleus. EMBO J 18:1953–1965

Bucci M, Wente S (1997) In vivo dynamics of nuclear pore complexes in yeast. J Cell Biol 136:1185–1199

Bucci M, Wente SR (1998) A novel fluorescence-based genetic strategy identifies mutants of *Saccharomyces cerevisiae* defective for nuclear pore complex assembly. Mol Biol Cell 9:2439–2461

Buendia B, Santa-Maria A, Courvalin JC (1999) Caspase-dependent proteolysis of integral and peripheral proteins of nuclear membranes and nuclear pore complex proteins during apoptosis. J Cell Sci 112:1743–1753

Chial HJ, Rout MP, Giddings Jr TH, Winey M (1998) *Saccharomyces cerevisiae* Ndc1p is a shared component of nuclear pore complexes and spindle pole bodies. J Cell Biol 143:1789–1800

Chook YM, Blobel G (1999) Structure of the nuclear transport complex karyopherin-β2-Ran.GppNHp. Nature 399:230–237

Collas P (1998) Nuclear envelope disassembly in mitotic extracts requires functional nuclear pores and a nuclear lamina. J Cell Sci 111:1293–1303

Conti E, Uy M, Leighton L, Blobel G, Kuriyan J (1998) Crystallographic analysis of the recognition of a nuclear localization signal by the nuclear import factor karyopherin alpha. Cell 94:193–204

Cordes VK, Reidenbach S, Rackwitz HR, Franke WW (1997) Identification of protein p270/Tpr as a constitutive component of the nuclear pore complex-attached intranuclear filaments. J Cell Biol 136:515–529

Cordes VC, Hase ME, Müller L (1998) Molecular segments of protein Tpr that confer nuclear targeting and association with the nuclear pore complex. Exp Cell Res 245:43–56

Courvalin JC, Worman HJ (1997) Nuclear envelope protein autoantibodies in primary biliary cirrhosis. Semin Liver Dis 17:79–90

Del Priore V, Heath CV, Snay CA, Mac Millan A, Gorsch LC, Dagher S, Cole CN (1997) A structure/function analysis of Rat7p/Nup159, an essential nucleoporin of *Saccharomyces cerevisiae*. J Cell Sci 110:2987–2999

Dockendorff TC, Heath CV, Goldstein AL, Snay CA, Cole CN (1997) C-terminal truncations of the yeast nucleoporin Nup145p produce a rapid temperature-conditional mRNA export defect and alternations to nuclear structure. Mol Cell Biol 17:906–920

Doye V, Hurt E (1997) From nucleoporins to nuclear pore complexes. Curr Opin Cell Biol 9:401–411

Emtage JLT, Bucci M, Watkins JL, Wente SR (1997) Defining the essential functional regions of the nucleoporin Nup145p. J Cell Sci 110:911–925

Enarson P, Enarson M, Bastos R, Burke B (1998) Amino-terminal sequences that direct nucleoporin Nup153 to the inner surface of the nuclear envelope. Chromosoma 107:228–236

Engel A, Lyubchenko Y, Muller D (1999) Atomic force microscopy: a powerful tool to observe biomolecules at work. Trends Cell Biol 9:77–80

Fabre E, Hurt E (1997) Yeast genetics to dissect the nuclear pore complex and nucleo-cytoplasmic trafficking. Ann Rev Genet 31:277–313

Fahrenkrog B, Hurt EC, Aebi U, Panté N (1998) Molecular architecture of the yeast nuclear pore complex: localization of Nsp1p subcomplexes. J Cell Biol 143:577–588

Fahrenkrog B, Aris JP, Hurt EC, Panté N, Aebi U (2000a) Comparative localization of protein A tagged and endogenous yeast nuclear pore complex proteins by immunoelectron microscopy. J Struct Biol 129:295–305

Fahrenkrog B, Hübner W, Mandinova A, Panté N, Keller W, Aebi U (2000b) The yeast nucleoporin Nup53p specifically interacts with Nic96p and is directly involved in nuclear protein import. Mol Biol Cell 11:3885–3896

Farjot G, Sergant A, Mikaélian I (1999) A new nucleoporin-like protein interacts with both HIV-1 Rev nuclear export signal and CRM-1. J Biol Chem 274:17309–17317

Favreau C, Worman HJ, Wozniak RW, Frappier T, Courvalin JC (1996) Cell cycle-dependent phosphorylation of nucleoporins and nuclear pore membrane protein Gp210. Biochemistry 35:8035–8044

Feldherr C, Akin D, Moore MS (1998) The nuclear import factor p10 regulates the functional size of the nuclear pore complex during oogenesis. J Cell Sci 111:1889–1896

Floer M, Blobel G (1999) Putative reaction intermediates in Crm1-mediated nuclear protein export. J Biol Chem 274:16279–16286

Fontoura BMA, Blobel G, Matunis MJ (1999) A conserved biogenesis pathway for nucleoporins: proteolytic processing of a 186-kilodalton precursor generates Nup98 and the novel nucleoporin, Nup96. J Cell Biol 144:1097–1112

Fornerod M, van Deursen J, van Baal S, Reynolds A, Davis D, Murti KG, Fransen J, Grosveld G (1997) The human homologue of yeast Crm1 is in a dynamic subcomplex with CAN/Nup214 and a novel nuclear pore component Nup88. EMBO J 16:807–816

Fouchier RAM, Meyer BE, Simon JHM, Fischer U, Albright AV, Gonzalez-Scarrano F, Malim MH (1998) Interaction of the Human Immundeficiency Virus Type1 Vpr protein with the nuclear pore complex. J Virol 72:6004–6013

Gigliotti S, Callaini G, Andone S, Riparbelli MG, Pernas-Alonso R, Hoffmann G, Grazani F, Malva C (1998) Nup154, a new *Drosophila* gene essential for male and female gametogenesis is related to the Nup155 vertebrate nucleoporin gene. J Cell Biol 142:1195–1207

Grandi P, Dang T, Panté N, Shevchenko A, Mann M, Forbes D, Hurt E (1997) Nup93, a vertebrate homologue of yeast Nic96p, forms a complex with a novel 205-kDa protein and is required for correct nuclear pore assembly. Mol Biol Cell 8:2017–2038

Grüter P, Taberno C, von Kobbe C, Schmitt C, Saavedra C, Bachi A, Wilm M, Felber BK, Izaurralde E (1998) TAP, the human homolog of Mex67p, mediates CTE-dependent RNA export from the nucleus. Mol Cell 1:649–659

Ho AK, Raczniak GA, Ives EB, Wente SR (1998) The integral membrane protein Snl1p is genetically linked to yeast nuclear pore complex function. Mol Biol Cell 9:355–373

Hu T, Gerace L (1998) cDNA cloning and analysis of the expression of nucleoporin p45. Gene 221:245–253

Hurwitz ME, Strambio-de-Castillia C, Blobel G (1998) Two yeast nuclear pore complex proteins involved in mRNA export of a cytoplasmically oriented subcomplex. Proc Natl Acad Sci USA 95:11241–11245

Hussey DJ, Nicola M, Moore S, Peters GB, Dobrovic A (1999) The (4;11)(q21;p15) translocation fuses the *NUP98* and *RAP1GDS1* genes and is recurrent in T-cell acute lymphocytic leukemia. Blood 94:2072–2079

Ikeda T, Ikeda K, Sasaki K, Kawakami K, Takahara J (1999) The inv(11)(p15q22) chromosome translocation of therapy-related myelodysplasia with NUP98-DDX10 and DDX10-NUP98 fusion transcripts. Int J Hematol 69:160–164

Iovine MK, Wente SR (1997) A nuclear export signal in Kap95p is required for both recycling the import factor and the interaction with the nucleoporin GLFG repeat regions of Nup116p and Nup100p. J Cell Biol 137:797–811

Izaurralde E, Kann M, Panté N, Sodeik B, Hohn T (1999) Viruses, microorganisms and scientists meet the nuclear pore. EMBO J 18:289–296

Izaurralde E, Adam SA (1998) Transport of macromolecules between the nucleus and the cytoplasm. RNA 4:351–364

Jarnik M, Aebi U (1991) Towards a 3-D model of the nuclear pore complex. J Struct Biol 107:291–308

Jenkins Y, McEntee M, Weis K, Greene WC (1998) Characterization of HIV-1 Vpr nuclear import: analysis of signals and pathways. J Cell Biol 143:875–885

Kang Y, Cullen BR (1999) The human TAP protein is a nuclear mRNA export factor that contains novel RNA-binding and nucleo-cytoplasmic transport sequences. Genes Dev 13:1126–1139

Kasamatsu H, Nakanishi A (1998) How do animal DNA viruses get to the nucleus? Annu Rev Microbiol 52:627–686

Kasper LH, Brindle PK, Schnabel CA, Pritchard CEJ, Cleary ML, van Deursen JMA (1999) CREB binding protein interacts with nucleoporin-specific FG repeats that activate transcription and mediate NUP98-HOAX9 oncogenicity. Mol Cell Biol 19:764–776

Katahira J, Sträßer K, Podtelejnikov A, Mann M, Jung JU, Hurt E (1999) The Mex67p-mediated nuclear mRNA export pathway is conserved from yeast to human. EMBO J 18:2593–2609

Kinoshita H, Omagari K, Whittingham S, Kato Y, Ishibashi H, Sugi K, Yano M, Kohno S, Nakanuma Y, Penner E, Wesierska-Gadek J, Reynoso-Paz S, Gershwin ME, Anderson J, Jois JA, Mackay IR (1999) Autoimmune cholangitis and primary biliary cirrhosis – an autoimmune enigma. Liver 19:122–128

Kiseleva E, Goldberg MW, Allen TD, Akey CW (1998) Active nuclear pore complexes in Chironomous: visualization of transporter configurations related to mRNP export. J Cell Sci 111:223–236

Kosova B, Panté N, Rollenhagen C, Hurt E (1999) Nup192p is a conserved nculeoporin with a preferential location at the inner site of the nuclear membrane. J Biol Chem 274:22646–22651

Kosova B, Panté N, Rollenhagen C, Podtelejnikov A, Mann M, Aebi U, Hurt E (2000) Mlp2p, a component of nuclear pore attached intranuclear filaments, associates with Nic96p. J Biol Chem 275:343–350

Kraemer DM, Strambio-de-Castilla C, Blobel G, Rout MP (1995) The essential yeast nucleoporin NUP159 is located on the cytoplasmic side of the nuclear pore complex and serves in karyopherin-mediated binding of transport substrate. J Biol Chem 270:19017–19021

Kwong YL, Pang A (1999) Low frequency of rearrangements of the homeobox gene HOXA9/t(7;11) in adult acute myeloid leukemia. Genes Chromosomes Cancer 25:70–74

Macaulay C, Meier E, Forbes DJ (1995) Differential mitotic phosphorylation of proteins of the nuclear pore complex. J Biol Chem 270:254–262

Malim MH, Hauber J, Le SY, Maizel J, Cullen BR (1989) The HIV-rev trans-activator acts through a structured target sequence to activate nuclear export of unspliced viral mRNA. Nature 338:254–257

Marelli M, Aitchinson JD, Wozniak RW (1998) Specific binding of the karyopherin Kap121p to a subunit of the nuclear pore complex containing Nup53p, Nup59p, and Nup170p. J Cell Biol 143:1813–1830

Matsuoka Y, Takagi M, Ban T, Miyazaki M, Yamamoto T, Kondo Y, Yoneda Y (1999) Identification and characterization of nuclear pore subcomplexes in mitotic extract of human somatic cells. Biochem Biophys Res Commun 254:417–423

Mattaj IW, Englmeier L (1998) Nucleo-cytoplasmic transport: the soluble phase. Annu Rev Biochem 67:265–306

Nakamura T, Largaespada DA, Lee MP, Johnson LA, Ohyashiki K, Toyama K, Chen SJ, Willman CL, Chen IM, Feinberg AP, Jenkins NA, Copeland NG, Shaugnessy Jr JD (1996) Fusion of the nucleoporin gene *NUP98* to *HOXA9* by the chromosome translocation t(7;11)(p15;p15) in human myeloid leukemia. Nature Genetics 12:154–158

Nakamura T, Yamazaki Y, Hatano Y, Miura I (1999) *NUP98* is fused to *PMX1* homeobox gene in human acute myelogenous leukemia with chromosome translocation t(1;11)(q23;p15). Blood 94:741–747

Nakielny S, Shaikh S, Burke B, Dreyfuss G (1999) Nup153 is an M9-containing mobile nucleoporin with a novel Ran-binding domain. EMBO J 18:1982–1995

Nickowitz RE, Worman HJ (1993) Autoantibodies from patients with primary biliary cirrhosis recognize a restricted region within the cytoplasmic tail of nuclear pore membrane protein gp210. J Exp Med 178:2237–2242

Nishiyama M, Arai Y, Tsunematsu Y, Kobayashi H, Asami K, Yabe M, Kato S, Oda M, Eguchi H, Ohki M, Kaneko Y (1999) 11p15 translocations involving the NUP98 gene in childhood therapy-related acute myeloid leukemia/myelodysplastic syndrome. Genes Chromosomes Cancer 26:215–220

Ohno M, Fornerod M, Mattaj IW (1998) Nucleo-cytoplasmic transport: the last 200 nanometers. Cell 92:327–336

O'Neill RE, Talon J, Palese P (1998) The influenza virus NEP (NS2 protein) mediates the nuclear export of viral ribonucleoproteins. EMBO J 17:288–296

Otero GC, Harris ME, Donello JE, Hope TJ (1998) Leptomycin B inhibits equine infectious anemia virus Rev and feline immunodeficiency virus Rev function but not the function of hepatitis B virus post-transcriptional regulatory element. J Virol 72:7593–7597

Panté N, Bastos R, McMorrow I, Burke B, Aebi U (1994) Interactions and three-dimensional localization of a group of nuclear pore complex proteins. J Cell Biol 129:925–937

Panté N, Aebi U (1996a) Molecular dissection of the nuclear pore complex. Crit Rev Biochem Mol Biol 31:153–199

Panté N, Aebi U (1996b) Toward the Molecular Dissection of Protein Import into Nuclei. Curr Opin Cell Biol 8:397–406

Perez-Terzic C, Gacy AM, Bortolon R, Dzeja PP, Puceat M, Jaconi M, Prendergast FG, Terzic A (1999) Structural plasticity of the cardiac nuclear pore complex in response to regulators of nuclear import. Circ Res 84:1292–1301

Popov S, Rexach M, Zybarth G, Reiling N, Lee MA, Ratner L, McLane C, Moore MS, Blobel G, Bukrinsky M (1998) Viral protein R regulates nuclear import of the HIV-1 pre-integration complex. EMBO J 17:909–917

Popov S, Rexach M, Ratner L, Blobel G, Bukrinsky M (1998a) Viral protein R regulates docking of the HIV-1 preintegration complex to the nuclear pore complex. J Biol Chem 273:13347–13352

Powers M, Macaulay C, Masiarz FR, Forbes DJ (1995) Reconstituted nuclei depleted of a vertebrate GLFG nuclear pore protein, p97, import but are defective in nuclear growth and replication. J Cell Biol 128:721–736

Powers M, Forbes DJ, Dahlberg JE, Lund E (1997) The vertebrate GLFG nucleoporin, Nup98, is an essential component of multiple RNA export pathways. J Cell Biol 136:241–250

Radu A, Moore MS, Blobel G (1995) The peptide repeat domain of nucleoporin Nup98 functions as docking site in transport across the nuclear pore complex. Cell 81:215–222

Rakowska A, Danker T, Schneider SW, Oberleithner H (1998). ATP-induced shape changes of nuclear pores visualized with the atomic force microscope. J Membrane Biol 163:129–136

Raza-Egilmez SZ, Jani-Sait SN, Grossi M, Higgins MJ, Shows TB, Aplan PD (1998) NUP98-HOXD13 gene fusion in therapy-related acute myelogenous leukemia. Cancer Research 58:4269–4273

Rexach M, Blobel G (1995) Protein import into nuclei: Association and dissociation reactions involving transport substrate, transport factors and nucleoporins. Cell 83:683–692

Rosenblum JS, Blobel G (1999) Autoproteolysis in nucleoporin biogenesis. Proc Natl Acad Sci USA 96:11370–11375

Roulston A, Marcellus RC, Branton PE (1999) Viruses and apoptosis. Annu Rev Microbiol 53:577–628

Rout MP, Blobel G (1993) Isolation of the yeast nuclear pore complex. J Cell Biol 109:2641–2652

Saavedra C, Felber BK, Izaurralde E (1997) The simian retrovirus-1 constitutive transport element, unlike the HIV-1 RRE, utilises factors required for the export of cellular RNAs. Curr Biol 7:619–628

Santos-Rosa H, Moreno H, Simos G, Segref A, Fahrenkrog B, Panté N, Hurt E (1998) Nuclear mRNA export requires complex formation between Mex67p and Mtr2p at the nuclear pores. Mol Cell Biol 18:6826–6838

Scheffzek K, Klebe C, Fritz-Wolf K, Kabsch W, Wittinghofer A (1995) Crystal structure of the nuclear Ras-related protein Ran in its GDP-bound form. Nature 374:378–381

Seedorf M, Damelin M, Kahana J, Taura T, Silver PA (1999) Interactions between a nuclear transporter and a subset of nuclear pore complex proteins depend on Ran GTPase. Mol Cell Biol 19:1547–1557

Segref A, Sharma K, Doye V, Hellwig A, Huber J, Hurt EC (1996) Mex67p, which is an essential factor for nuclear mRNA export, binds to both poly(A)$^+$ RNA and nuclear pores. EMBO J 16:3256–3271

Siniossoglou S, Santos-Rosa H, Rappsilber J, Mann M, Hurt E (1999) A novel complex of membrane proteins required for formation of a spherical nucleus. EMBO J 17:6449–6464

Shah S, Tugendreich S, Forbes D (1998) Major binding sites for the nuclear import receptor are the integral nucleoporin Nup153 and the adjacent nuclear filament protein Tpr. J Cell Biol 141:31–49

Shah S, Forbes DJ (1998) Separate nuclear import pathways converge on the nucleoporin Nup153 and can be dissected with dominant-negative inhibitors. Curr Biol 8:1376–1386

Söderqvist H, Imreh G, Kihlmark M, Linnmann C, Ringertz N, Hallberg E (1997) Intracellular distribution of an integral nuclear pore membrane protein fused to green fluorescent protein. Eur J Biochem 250:808–813

Stewart M, Kent HM, McCoy AJ (1998) Structural basis for molecular recognition between nuclear transport factor 2 (NTF2) and the GDP-bound form of the Ras-family GTPase Ran. J Mol Biol 277:635–646

Stochaj U, Héjazi M, Belhumeur P (1998) The small GTPase Gsp1p binds to the repeat domain of the nucleoporin Nsp1p. Biochem J 330:412–427

Stoffler D, Fahrenkrog B, Aebi U (1999a) The nuclear pore complex: from molecular architecture to functional dynamics. Curr Opin Cell Biol 11:391–401

Stoffler D, Goldie KN, Aebi U (1999b) Calcium-mediated structural changes of native nuclear pore complexes monitored by time-lapse atomic force microscopy. J Mol Biol 287:741–752

Strahm Y, Fahrenkrog B, Zenklusen D, Rycher E, Kantor J, Rosbach M, Stutz F (1999) The RNA export factor Gle1p is located on the cytoplasmic fibrils of the NPC and physically interacts with the FG-nucleoporin Rip1p, the DEAD-box protein Rat8p/Dbp5p and a new protein Ymr255p. EMBO J 18:5761–5777

Strambio-de-Castillia C, Blobel G, Rout MP (1999) Proteins connecting the nuclear pore complex with the nuclear interior. J Cell Biol 144:839–855

Starmbio-de-Castillia C, Rout MP (1999) TAPping into transport. Nature Cell Biol 1:E31–E33

Stutz F, Neville M, Rosbash M (1995) Identification of a novel nuclear pore-associated protein as a functional target for HIV-1 Rev protein in yeast. Cell 82:495–506

Stutz F, Rosbash M (1998) Nuclear RNA export. Genes Dev 12:3303–3319

Teixeira MT, Siniossoglou S, Podtelejnikov S, Bénichou JC, Mann M, Dujon B, Hurt E, Fabre E (1997) Two functionally distinct domains generated by in vivo cleavage of Nup145p: a novel biogenesis pathway for nucleoporins. EMBO J 16:5086–5097

Theodoropoulos PA, Polioudaki H, Koulentaki M, Kouroumalis E, Georgatos SD (1999) PBC68: a nuclear pore complex protein that associates reversibly with the mitotic spindle. J Cell Sci 112:3049–3059

Van Deursen J, Boer J, Kasper L, Grosveld G (1996) G2 arrest and impaired nucleo-cytoplasmic transport in mouse embryos lacking the proto-oncogene CAN/Nup214. EMBO J 15:5574–5583

Vetter IR, Arndt A, Kutay U, Görlich D, Wittinghofer A (1999a) Structural view of the Ran-Importin beta interaction at 2.3 Å resolution. Cell 97:635–646

Vetter IR, Nowak C, Nishimoto T, Kuhlmann J, Wittinghofer A (1999b) Structure of a Ran-binding domain complexed with Ran bound to a GTP analogue: implications for nuclear transport. Nature 398:39–46

Vodicka MA, Koepp DM, Silver PA, Emermann M (1998) HIV-1Vpr interacts with the nuclear transport pathway to promote macrophage infection. Genes Dev 12:175–185

Wang H, Clapham DE (1999) Conformational changes of the in situ nuclear pore complex. Biophys J 77:241–247

Wertz IE, Hanley MR (1996) Diverse molecular provocation of programmed cell death. Trends Biochem Sci 21:359–364

West RR, Vaisberg EV, Ding R, Nurse P, McIntosh JR (1998) $cut11^+$: a gene required for cell cycle-dependent spindle pole body anchoring in the nuclear envelope and bipolar spindle formation in *Schizosaccharomyces pombe*. Mol Biol Cell 9:2839–2855

Whittaker GR, Helenius A (1998) Nuclear import and export of viruses and virus genome. Virology 246:1–23

Yan C, Leibowitz N, Mélèse T (1997) A role for the divergent actin gene, *ACT2*, in nuclear pore structure and function. EMBO J 16:3572–3586

Yang Q, Rout MP, Akey C (1998) Three-dimensional architecture of the isolated yeast nuclear pore complex: functional and evolutionary implications. Mol Cell 1:223–234

Yaseen NR, Blobel G (1997) Cloning and characteriziation of human karyopherin β3. Proc Natl Acad Sci USA 94:4451–4456

Yoon JH, Whalen WA, Bharathi A, Shen R, Dhar R (1997) Npp106p, a *Schizosaccharomyces pombe* nucleoporin similar to *Saccharomyces cerevisiae* Nic96p, functionally interacts with Rae1p in mRNA export. Mol Cell Biol 17:7047–7060

Zimowska G, Aris JP, Paddy MR (1997) A *Drosophila* Tpr protein homolog is localized both in the extrachromosomal channel network and to nuclear pore complexes. J Cell Sci 110:927–944

Zhang X, Huanming Y, Corydon MJ, Zhang X, Pedersen S, Korenberg JR, Chen XN, Laporte J, Gregersen N, Niebuhr E, Liu G, Bolund L (1999a) Localization of a human nucleoporin 155 gene (NUP155) to the 5p13 region and cloning of its cDNA. Genomics 57:144–151

Zhang C, Hughes M, Clarke PR (1999b) Ran-GTP stabilises microtubule asters and inhibits nuclear assembly in *Xenopus* egg extracts. J Cell Sci 112:2453–2461

Zolotukhin A, Felber BK (1999) Nucleoporins Nup98 and Nup214 participate in nuclear export of human immunodeficiency virus type 1 Rev. J Virol 73:120–127

Note added in proof. Recently, by a proteomics approach all yeast nucleoporins have been identified and immunolocalized.

Rout MP, Aitchison SD, Suprapto A, Hjertaas K, Zhao Y, Chait BT (2000) The yeast nuclear pore complex: composition, architecture, and transport mechanism. J Cell Biol 148:635–651

Methods and Assays to Investigate Nuclear Export

R.H. STAUBER

1	Introduction 119
2	In Vivo Assays 120
2.1	Visualizing Nucleo-Cytoplasmic Trafficking by Drug Treatment 120
2.2	Heterokaryon Analysis to Investigate Nucleo-Cytoplasmic Shuttling 121
2.3	Microinjection of Recombinant Proteins to Study Export 123
2.4	The Rapamycin-Mediated Linkage System to Study Nuclear Export 124
3	In Vitro Assays 124
3.1	Using Fusions Between a HIV-1 Rev-GFP Hybrid and the Hormone-Responsive Element from the Glucocorticoid Receptor to Study Export 125
3.2	Exploiting the Biotin–Streptavidin Interaction to Set Up Export Assays 125
3.3	The NFAT-System to Study the Requirements for Export 125
3.4	Optical Single-Transporter Recording to Study Nuclear Export 126
4	Concluding Remarks 126
References	127

1 Introduction

A hallmark of eukaryotic cells is their spatial and functional separation into the nucleus and the cytoplasm by the nuclear envelope. Although this separation introduces a potent and sophisticated level of regulation not existing in prokaryotes, it also requires a highly effective and selective transport machinery. All known transport between the nucleus and the cytoplasm occurs through the nuclear pore complex (NPC) (for reviews, see FABRE and HURT 1997; GANT et al. 1998; PANTÈ and AEBI 1996). Theoretically, proteins with masses <40kDa can enter and leave the nucleus by passive diffusion. However, even most of the smaller proteins and nucleic acids appear to be transported by signal-mediated pathways, probably because signal-mediated trafficking is more efficient and more amenable to specific regulation than diffusion. The most prevalent nuclear export signals (NESs) found consist of a short leucine-rich stretch of amino acids in

---

Institute for Clinical and Molecular Virology, University of Erlangen-Nürnberg, Schlossgarten 4, 91054 Erlangen, Germany

which the leucine residues are critical for function (for review, see GÖRLICH and KUTAY 1999; MATTAJ and ENGLMEIER 1998 and references therein). Leucine-rich NESs have been identified in an increasing number of cellular and viral proteins executing quite heterologous biological functions. As extensively reviewed in the individual chapters of this book, viruses efficiently exploit the cellular transport machinery in order to successfully parasitize the cell. This chapter will summarize existing in vitro and in vivo assay systems to study nuclear export in mammalian cells, focusing especially on novel experimental approaches. As a result of space limitation, important primary papers will not be quoted and we apologize to our colleagues for these omissions.

2 In Vivo Assays

2.1 Visualizing Nucleo-Cytoplasmic Trafficking by Drug Treatment

The first step in studying the requirements of nuclear export for a given protein is to demonstrate that the protein is indeed capable of nucleo-cytoplasmic shuttling. A simple approach is the use of chemical compounds that cause a change in the steady-state localization of the protein under investigation. Before and after drug treatment, the protein is visualized by indirect immunofluorescence or, in living cells, by using fusions to autofluorescent proteins (e.g., GFP, BFP, RFP) (MATZ et al. 1999; STAUBER et al. 1998; TSIEN 1998). For shuttle proteins displaying a predominantly cytoplasmic localization, the use of export inhibitors (e.g., Leptomycin B) will result in nuclear accumulation (Fig. 1A,B) (HEGER et al. 1999; KRÄTZER et al. 2000; TSIEN 1998). Leptomycin B (LMB) binds and inactivates the CRM1 export receptor, thereby blocking the nuclear export of proteins containing a leucine-rich NES (FORNEROD et al. 1997). Alternatively, transcription inhibitors have been used to demonstrate nucleo-cytoplasmic transport of several proteins. As reported, actinomycin D (ActD) or 5,6-dichlororibofuranosylbenzimidazole (DRB) appear to block nuclear import of hnRNPA1 (PINOL-ROMA and DREYFUSS 1992) or the HIV-1 Rev protein (HENDERSON and ELEFTHERIOU 2000; MEYER and MALIM 1994; RICHARD et al. 1997) and affect nuclear export of the von Hippel-Lindau tumor suppressor protein (LEE et al. 1999) or the poly-A binding protein 1 (AFONINA et al. 1998). However, whether nucleo-cytoplasmic transport is directly inhibited or the effect is indirect by affecting cytoplasmic/nuclear retention is still a controversial issue. In the case of HIV-1 Rev (Fig. 1C,D), drug treatment appears to affect secondary binding sites in the nucleus (e.g., 5S RNA at the nucleolus which binds Rev) (LEE et al. 1999; STAUBER et al. 1998) and therefore changes the steady-state localization of the proteins rather than interfering directly with transport (D'AGOSTINO et al. 1995; HEGER et al. 1998).

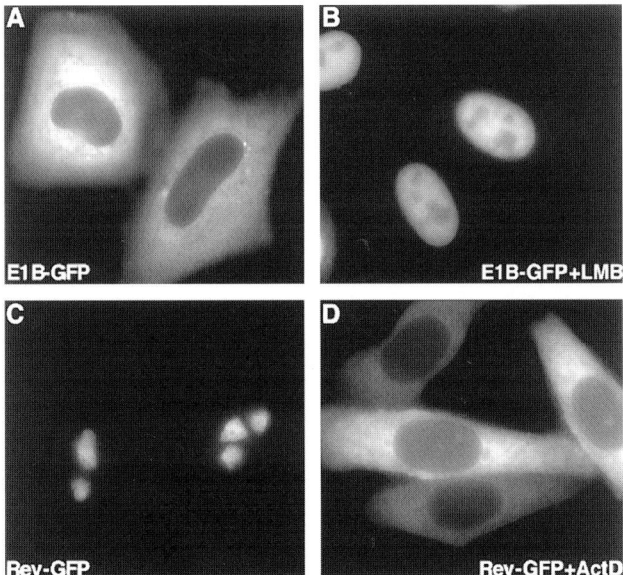

Fig. 1A–D. Drug treatment to demonstrate nucleo-cytoplasmic trafficking. Cells were transfected with the indicated plasmids and analyzed by fluorescence microscopy 16h later. In living HeLa cells the adenovirus type 5 E1B-55K-GFP hybrid localized predominantly to the cytoplasm (**A**) and accumulated in the nucleus following leptomycin B treatment (**B**). Nucleolar HIV-1 Rev-GFP (**C**) accumulated in the cytoplasm following treatment with ActD (**D**)

2.2 Heterokaryon Analysis to Investigate Nucleo-Cytoplasmic Shuttling

For nuclear proteins not responding to drug treatment, the heterokaryon fusion assay represents an efficient in vivo approach (AFONINA et al. 1998; IZAURRALDE et al. 1997). In a standard fusion assay, cells expressing the protein of interest (e.g., stable cell lines or by transient expression) are mixed with non-expressing cells. After polyethylene glycol (PEG) treatment the plasma membranes of the individual cells start to fuse, resulting in a multinuclear syncytium. If the protein of interest is constantly exported from and imported into the donor nucleus, it will be imported also into the acceptor nuclei over time (Fig. 2A). The use of autofluorescent proteins (AFP)-tagged proteins allows monitoring of shuttling in living cells and comparison of the transport kinetics between different proteins. However, one has to keep in mind that fusion assays represent a mixture of nuclear export (from the donor nucleus) and nuclear import (into the donor and acceptor nucleus). To facilitate the discrimination of donor and acceptor nuclei, mouse cells are often used as acceptor cells because the nuclei of human and mouse cells can easily be distinguished by staining with Hoechst dye. Because prolonged or extensive treatment with PEG can also cause disruption of the nuclear envelope, the experimental conditions have to be controlled by including a non-shuttling protein as a negative control (Fig. 2B,C).

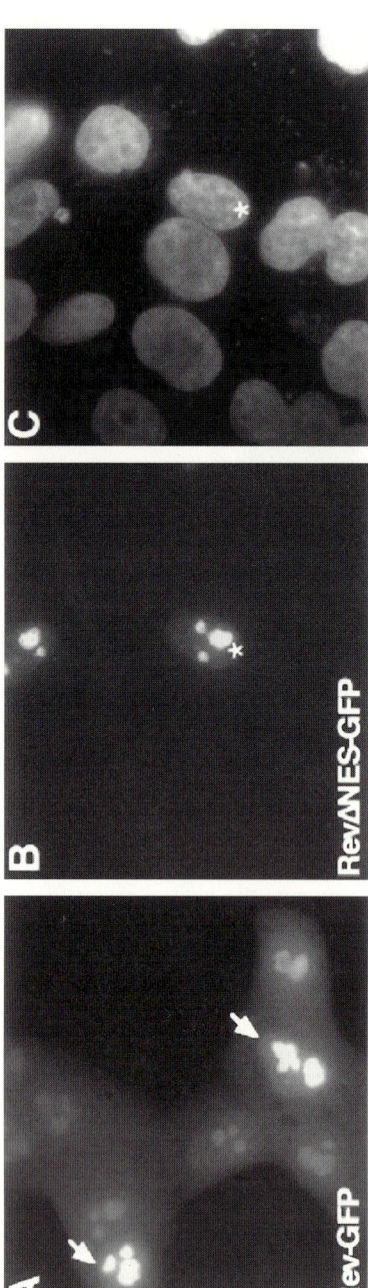

Fig. 2A–C. Cell fusion assay to demonstrate nucleo-cytoplasmic shuttling of the HIV-1 Rev protein. Cells were transfected with the indicated plasmids and 16h later treated with PEG for 1min and analyzed by fluorescence microscopy. Whereas Rev-GFP was efficiently exported and accumulated in the nuclei of the acceptor cells, RevM10-GFP harboring an inactive NES remained in the nucleus of the donor cells. **A,B** Rev-GFP or RevΔNES-GFP, respectively, 15min post-fusion. **C** Staining of surrounding nuclei in **B** with Hoechst dye to visualize the presence of acceptor nuclei. *Arrows* indicate donor nuclei. *Asterisks* mark the RevΔNES-GFP-expressing cell in **B** and **C**

2.3 Microinjection of Recombinant Proteins to Study Export

Capillary microinjection of recombinant proteins has proven to be an efficient approach to investigate directly nuclear export independent of import or vice versa (for technical details, see CID-ARREGUI and GARCIA-CARRANCÁ and references therein). A limitation is often the efficient expression and purification of large or toxic proteins in sufficient amounts to study export of the full-length proteins. Thus, this technique has been mostly used to identify and characterize in detail nuclear export signals (BOGERD et al. 1999; WEN et al. 1995). NESs are either expressed as a fusion with a heterologous protein (e.g., GST) or NES peptides are conjugated to bulky, fluorescently labeled carrier proteins (e.g., BSA) to avoid passive intracellular diffusion. Subsequently, the substrates are injected into the nuclei of somatic cells and export of the NES-fusions is monitored by immunostaining or direct fluorescence. A more elegant approach represents the use of NESs linked to a chimeric protein composed of GST fused to GFP, which allows recording of real-time kinetics of export (ROSORIUS et al. 1998). Transport of the stable and highly fluorescent substrates can be observed directly by fluorescence microscopy in living cells following microinjection (Fig. 3). In addition, the size of GST-GFP (~54kDa, as a monomer) prevents intracellular passive diffusion and the combination of GST/GFP-tagging allows control of protein expression and monitoring of protein purification.

Microinjection also allows the introduction of specific inhibitors (e.g., antibodies, drugs, lectins) together with the export substrates in order to analyze their effects on individual RNA or protein export pathways (ELFGANG et al. 1999; ULLMAN et al. 1999).

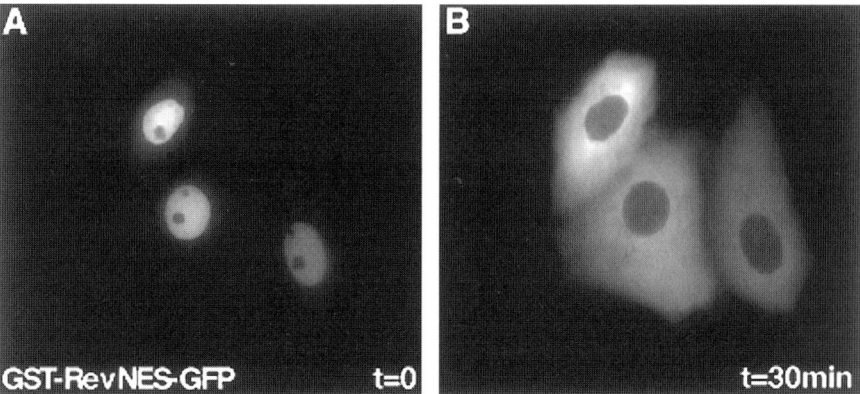

Fig. 3A,B. Nuclear export of a GST-RevNES-GFP hybrid in live cells. Purified GST-RevNES-GFP was microinjected into the nucleus of Vero cells and nuclear export was monitored directly by fluorescence microscopy. **A** 0 min post-injection. **B** Nuclear export is completed after 30 min

2.4 The Rapamycin-Mediated Linkage System to Study Nuclear Export

An inducible system was described by KLEMM et al. (1997). In this assay, a small protein containing a potential NES is conditionally linked to a protein localized in the nucleus by virtue of a NLS. The conditional linkage is mediated by rapamycin, a membrane-permeable macrolide that binds the immunophilin FK506-binding protein-12 (FKBP12). The FKBP12-rapamycin (FR) complex interacts with the FR-binding (FRB) domain of the FR-associated protein (FRAP). Two chimeric proteins, one containing FKBP12 and the other containing the FRB domain, can thus be linked in cells by the addition of rapamycin. Functional NESs are identified by their ability to direct export of the complex. In the experimental set-up described, a fusion protein containing the Gal4 DNA-binding domain NLS, linked to three repeated FKBP12 domains and marked with a FLAG epitope tag (Gal4-FKBP), served as the nuclear prey. The Rev NES fused to the FRB domain and marked with a hemagglutinin epitope (FRB-Rev) could passively diffuse into the nucleus and was used as the bait. Coexpression of both fusion proteins resulted in the cytoplasmic translocation of the prey induced by the addition of rapamycin and was visualized by indirect immunofluorescence using the α-FLAG antibody. This system can also be tailored for the identification of weak NESs by varying the number of FKBP12 NES-binding domains in the nuclear bait.

3 In Vitro Assays

In vivo export assays are important for the initial identification and characterization of export signals and pathways. However, to ultimately understand the molecular mechanism of export, in vitro assays that allow reconstitution of export from defined factors are required. The principle of the existing in vitro assays is to localize the export substrate into the nucleus, to deplete the cells of export factors (mostly by permeabilization of the plasma membrane using digitonin) and to reconstitute export by the addition of biochemically more or less well-defined components. In contrast to in vitro import assays, in which the nuclei are simply incubated with the import substrate, export assays require an initial nuclear-loading phase. Since the steady state localization of a protein is the net result of nuclear/cytoplasmic retention vs nuclear export, the success of the in vitro system often depends on choosing the right combination of nuclear import and export signals. This has hampered so far the development of an universal system to study the export requirements for a variety of shuttle proteins.

In the subsequent section, the principals of the reported systems are briefly described and the potential applications and limitations are discussed. This should assist the reader in choosing the most appropriate assay system.

3.1 Using Fusions Between a HIV-1 Rev-GFP Hybrid and the Hormone-Responsive Element from the Glucocorticoid Receptor to Study Export

The system described by LOVE et al. (1998) uses a GFP-labeled, hormone-inducible Rev chimeric protein (Rev-GR-GFP) that allows control over the nuclear import and export signals of Rev. In the absence of steroids, the Rev-GR-GFP protein localizes in the cytoplasm, most likely by the interaction with heat shock proteins (MADAN and DEFRANCO 1993). Addition of steroids triggers nuclear import and the Rev-hybrid accumulates in the nucleus/nucleolus. Following permeabilization by digitonin, export, i.e., the loss of nuclear fluorescence, can be induced by the addition of cytosol. The major advantage of this system compared to cells transiently expressing Rev-GFP is that stable Rev-GR-GFP-expressing cell lines are available to study transport. This system has been optimized to study Rev export, therefore, it has to be tested for the analysis of heterologous export signals. Since NESs differ in their potential to promote export (HENDERSON and ELEFTHERIOU 2000; R.H. Stauber, unpublished observation), weaker signals might not be able to overcome the nuclear/nucleolar retention mediated by the Rev NLS. On the other hand, stronger export signals might result in a cytoplasmic steady-state localization of the Rev hybrid even in the presence of steroids.

3.2 Exploiting the Biotin–Streptavidin Interaction to Set Up Export Assays

In order to investigate the requirements for nuclear export of the protein kinase inhibitor (PKI), HOLASKA and colleagues (HOLASKA and PASCHAL 1998) developed an elegant assay system. The relatively small size (75 amino acids) of biotinylated recombinant PKI (bPKI) allows it to rapidly diffuse into the nucleus, where it binds to a fluorescently labeled streptavidin-NLS substrate with high affinity and mediate its export to the cytoplasm. This system requires an initial import phase during which digitonin-permeabilized nuclei are loaded with a recombinant fluorescent streptavidin-NLS fusion protein (FITC-STV-NLS) in the presence of cytosol. After the initial import phase, the permeabilized cells are washed and bPKI is added, which diffuses into the nuclei and binds to FITC-STV-NLS. The samples can now be incubated with buffer or cytosolic proteins and the levels of nuclear export is quantitated by measuring the loss of nuclear fluorescence over time.

3.3 The NFAT-System to Study the Requirements for Export

KEHLENBACH et al. (1998) developed a permeabilized cell assay to study export of the nuclear factor of activated T cells (NFAT). NFAT is a shuttle protein containing regulatable NLS as well as NES sequences. In resting cells, NFAT resides in the cytoplasm and import is triggered by elevated calcium levels activating the phosphatase calcineurin, leading to the dephosphorylation of NFAT and the exposure of the NLSs. Upon return of calcium to resting levels, NFAT is rapidly exported to the cytoplasm, requiring rephosphorylation of NFAT by the glycogen synthase kinase-3. Cell lines stably expressing a GFP-NFAT hybrid were treated with the calcium ionophore ionomycin, resulting in nuclear import of NFAT-GFP. Subsequently, the cells were permeabilized with digitonin, washed to remove the endogenous cytosol and preincubated to deplete nuclear export factors. To retain GFP-NFAT in the nucleus, lithium acetate, which inactivates glycogen synthase kinase-3, was present during the preincubation phase. Following a washing step to remove lithium, nuclear export of GFP-NFAT, i.e., the loss of nuclear fluorescence, can be observed microscopically or quantitated by flow cytometry after the addition of cytosol or nuclear extracts. The advantage of this rapid, quantitative in vitro assay is that a high number of export events can be recorded using flow-cytometry under various experimental conditions. However, so far the system has only been proven useful for the characterization of the NFAT export, but not as a general system to study export mediated by various NESs. As mentioned for the other export systems, this is due to the critical balance between nuclear import/retention and nuclear export. Replacing the NFAT-NES by the stronger Rev-NES resulted in a constitutive, cytoplasmic NFAT mutant that did not allow the initial nuclear loading phase by the addition of ionomycin (R.H. Stauber, unpublished observation).

3.4 Optical Single-Transporter Recording to Study Nuclear Export

The sophisticated in vitro system, optical single-transporter recording (OSTR), recently introduced by KEMINER et al. (1999), is obviating the initial nuclear loading phase and can be used to characterize the export requirements for a variety of NESs. By combining fluorescence microphotolysis, confocal scanning microscopy and membrane patching, OSTR is able to characterize the export capabilities of isolated *Xenopus* NPCs. Nuclear envelopes of *Xenopus* oocytes were attached to isoporous filters. Recombinant NLS or NES containing GFP-GST fusion proteins were applied to either the cytosolic or nuclear side of the pore, respectively, in the presence of egg or nuclear extracts. Transport of the fluorescent substrates out of the individual filter pores was subsequently measured by confocal scanning microscopy. Although the technical requirements for OSTR are demanding, this system allows quantitation of transport kinetics of individual NPCs. Furthermore, it can be applied to any transport signal and allows the investigator to freely manipulate the composition of the media on the cytoplasmic and nuclear side of the pore.

4 Concluding Remarks

The past few years have seen great progress in the characterization of nuclear import and export signals and their corresponding transport receptors. Sophisticated and efficient in vivo and in vitro systems have been developed in order to identify transport signals and to dissect the molecular mechanism regulating nuclear import and export. The challenge for the future will be to reconstitute export from biochemically defined components and to understand the orchestration of export by the concerted action of multiple cellular proteins and transport signals.

References

Afonina E, Stauber R, Pavlakis GN (1998) The human poly-A binding protein 1 shuttles between the nucleus and the cytoplasm. J Biol Chem 273:13015–13021

Bogerd HP, Benson RE, Truant R, Herold A, Phingbodhipakkiya M, Cullen BR (1999) Definition of a consensus transportin-specific nucleocytoplasmic transport signal. J Biol Chem 274:9771–9777

Cid-Arregui A, Garcia-Carrancá A (eds) (1998) Microinjection and transgenics. Springer, Berlin, Heidelberg New York

D'Agostino DM, Ciminale V, Pavlakis GN, Chieco-Bianchi L (1995) Intracellular trafficking of the human immunodeficiency virus type 1 Rev protein: involvement of continued rRNA synthesis in nuclear retention. AIDS Res Hum Retroviruses 11:1063–1072

Elfgang C, Rosorius O, Hofer L, Jakshe H, Hauber J, Bevec D (1999) Evidence for specific nucleoplasmic transport pathways used by leucine-rich nuclear export signals. Proc Natl Acad Sci USA 96:6229–6234

Fabre E, Hurt E (1997) Yeast genetics to dissect the nuclear pore complex and nucleocytoplasmic trafficking. Annu Rev Genet 31:277–313

Fornerod M, Ohno M, Yoshida M, Mattaj IW (1997) CRM1 is an export receptor for leucine-rich nuclear export signals [see comments]. Cell 90:1051–1060

Gant TM, Goldberg MW, Allen TD (1998) Nuclear envelope and nuclear pore assembly: analysis of assembly intermediates by electron microscopy. Curr Opin Cell Biol 10:409–415

Görlich D, Kutay U (1999) Transport between the cell nucleus and the cytoplasm. Annu Rev Cell Dev Biol 15:607–660

Heger P, Rosorius O, Hauber J, Stauber RH (1999) Titration of cellular export factors, but not heteromultimerization, is the molecular mechanism of *trans*-dominant HTLV-1 Rex mutants. Oncogene 18:4080–4090

Heger P, Rosorius O, Koch C, Casari G, Grassmann R, Hauber J (1998) Multimer-Formation is not essential for nuclear export of human T-cell leukemia virus type 1 Rex *trans*-activator protein. J Virol 72:8659–8668

Henderson BR, Eleftheriou A (2000) A comparison of the activity, sequence specificity, and CRM1-dependence of different nuclear export signals. Exp Cell Res 256:213–224

Holaska JM, Paschal BM (1998) A cytosolic activity distinct from Crm 1 mediates nuclear export of protein kinase inhibitor in permeabilized cells. Proc Natl Acad Sci USA 95:14739–14744

Izaurralde E, Jarmolowski A, Beisel C, Mattaj IW, Dreyfuss G, Fisher U (1997) A role for the M9 transport signal of hnRNP A1 in mRNA nuclear export. J Cell Biol 137:27–35

Kehlenbach RH, Dickmanns A, Gerace L (1998) Nucleocytoplasmic shuttling factors including Ran and CRM1 mediate nuclear export of NFAT in vitro. J Cell Biol 141:863–74

Keminer O, Siebrasse JP, Zerf K, Peters R (1999) Optical recording of signal-mediated protein transport through single nuclear pore complexes. Proc Natl Acad Sci USA 96:11842–11847

Klemm JD, Beals C, Crabree GR (1997) Rapid nuclear targeting of nuclear proteins to the cytoplasm. Current Biology 7:638–644

Krätzer FO, Rosrius O, Heger P, Hirschmann N, Dobner T, Hauber J, Stauber RH (2000) The adenovirus type 5 E1B-55K oncoprotein is a highly active shuttle protein and shuttling is independent of E4orf6, p53 and Mdm2. Oncogene 19:850–857

Lee S, Neumann M, Stearman R, Stuaber RH, Pause A, Pavlakis GN, Klausner RD (1999) Transcription dependent nuclear/cytoplasmic trafficking is required for the function of the von Hippel-Lindau tumor suppressor protein. Mol Cell Biol 19:1486–1497

Love DC, Sweitzer TD, Hanover JA (1998) Reconstitution of HIV-1 rev nuclear export: independent requirements for nuclear import and export. Proc Natl Acad Sci USA 95:10608–10613

Madan AP, DeFranco D (1993). Bidirectional transport of glucocorticoid receptor across the nuclear envelope. Proc Natl Acad Sci USA 90:3588–3592

Mattaj IW, Englmeier L (1998) Nucleocytoplasmic transport: the soluble phase. Annu Rev Biochem 67:265–306

Matz MV, Fradkov AF, Labas YA, Savitsky AP, Zaraisky AG, Markelov ML, Lukyanov SA (1999) Fluorescent proteins from nonbioluminiscent Antozoa species. Nat Biotech 17:969–973

Meyer BE, Malim MH (1994) The HIV-1 Rev *trans*-activator shuttles between the nucleus and the cytoplasm. Genes Dev 8:1538–1547

Pantè N, Aebi U (1996) Molecular dissection of the nuclear pore complex. Cri. Rev Biochem Mol Biol 31:153–199

Pinol-Roma S, Dreyfuss G (1992) Shuttling of pre-mRNA binding proteins between nucleus and cytoplasm. Nature 355:730–732

Richard N, Iacampo S, Cochrane A (1994) HIV-1 Rev is capable of shuttling between the nucleus and cytoplasm. Virology 204:123–131

Rosorius O, Heger P, Stelz G, Hirschmann N, Hauber J, Stauber RH (1999) Direct observation of nucleo-cytoplasmic transport by microinjection of GFP-tagged proteins in living cells. BioTechnique. 27:350–355

Stauber RH, Afonina E, Gulnik S, Erickson J, Pavlakis GN (1998) Analysis of intracellular trafficking and interactions of cytoplasmic HIV-1 Rev mutants in living cells. Virology 251:38–48

Stauber RH, Horie K, Carney P, Hudson EA, Tarasova NI, Gaitanaris GA, Pavlakis GN (1998) Development and applications of enhanced green fluorescent protein mutants. BioTechniques 24:462–471

Toyoshima F, Moriguchi T, Wada A, Fukuda M, Nishida E (1998) Nuclear export of cyclin B1 and its possible role in the DNA damage-induced G2 checkpoint. EMBO J 17:2728–2735

Tsien R (1998) The green fluorescent protein. Annu Rev Biochem 67:509–544

Ullman KS, Shah S, Powers MA, Forbes DJ (1999) The nucleoporin nup153 plays a critical role in multiple types of nuclear export. Mol Biol Cell 10:649–664

Wen W, Meinkoth JL, Tsien RY, Taylor SS (1995) Identification of a signal for rapid export of proteins from the nucleus. Cell 82:463–473

Subject Index

A
actinomycin D   120, 121
adeno-associated virus vector (rAAV)   41
adenovirus   121
– E1B-55K   121
apoptosis   43
ASF/SF2   35
atomic force microscopy (AFM)   105
avian sarcoma/leucemia (ASB/ALV) retroviruses   80

C
CAN (NUP214)   84
cell
– cycle   43, 44
– fusion assay   121, 122
constitutive transport element (CTE)   80–87
– ASV/ALV   80, 81
– MPMV   80–84
– 5′ splice site in   82, 83, 88
– SRV   80, 81, 84
CRM1   29, 30, 41, 42, 47, 48, 63, 81, 83
CRM1(p)   42
CTD, RNA polymerase II   87
cyclin A   46

D
dl338   45
dl1520   43, 45

E
E1A (see early region 1A)
E1B-AP5   38, 39, 42, 47, 48
eIF-4E   45
eIF-4F   45
eIF-5A   65
early region 1A (E1A)   36, 49
electron microscopic (EM)   95
export
– assays   120–126
– – in vitro   124–126
– – – biotin-streptavidin interaction export assay   125
– – – element from the glucocorticoid receptor to study export   125
– – – NFAT-system   125, 126
– – – optical single-transporter recording   126
– – in vivo   120–124
– – – heterokaryon analysis   121, 122
– – – microinjection of recombinant proteins   123
– – – rapamycin-mediated linkage system   124
– – – visualizing export by drug treatment   120, 121
– intron-containing RNA   78, 89
– mRNA   83, 84, 86, 87
– ribosomal subunits   83
– signals   85, 86
– 5S rRNA   83
– *Saccharomyces cerevisiae*   86, 87
– tRNA   83
– U snRNA   83, 89
– yeast   86, 87
exportin 1 (see CRM1)

G
G1 phase   44
genomic organization, retroviruses   79
GFP-GST   123, 126
Gle1   42
glutathione-*S*-transferase (GST)   39, 41, 48
green fluorescent protein (GFP)   41, 48
growth arrest   43
GST (see glutathione-*S*-transferase)
GTPase   39

H
heat shock (hs)   42
– protein 70   36
hepatitis B virus   4
herpes virus RNA export   1–17
Herpesvirus saimiri   85
heterokaryons   39, 40
HIV (see human immunodeficiency virus)

hnRNP A1 35
hnRNPs 39
hnRNP-U/SAF-A 39
host-range 43
hRip 42, 64
hs RNA 42
hsp70 42
HSV-1 5
human immunodeficiency virus (HIV) 78
- HIV-1 30, 31, 55
- HIV-1 Rev 29, 47, 120–125
human T-cell leukemia virus (HTLV) 78
- HTLV-1 39, 41, 55

I
ICP27 6–17
IG clusters 33, 38, 44, 47
IGs (interchromatin granules) 33
immunogold-EM 99
import receptor 85
importin/karyopherin-β superfamily 30
initiation factor e1F-4F 44, 45
interchromatin granules (IGs) 33
interferon-induced
- kinase DAI 45
- Mx-A 36, 42
intron retention 78, 79, 89
- CD44 89
- Id1 89
- Id3 89

K
Kr-bodies 32

L
L4 100-kDa 45
L5 mRNA 40, 41
leptomycin B (LMB) 41, 84, 120, 121
leucinerich NESs 41
LI-52/55 mRNA 36
LMB (*see* leptomycin B)

M
major late promoter (MLP) 27, 29, 34–37
major late transcription unit (MLTU) 27, 29, 34, 37
mammalian cells 81, 82
Mason-Pfizer monkey virus (MPMV) 79, 80
Mdm2 41
MEX67 86
Mex67p 86
microinjection 123
MLP (*see* major late promoter)
MLTU (*see* major late transcription unit)
mRNA translation 45

mutational analysis 81
- MPMV CTE 81, 82
- SRV CTE 81

N
Nbs (*see* nuclear bodies)
ND10 (*see* nuclear domains 10)
Nds (*see* nuclear dots)
NES (*see* nuclear export signal)
NFAT (*see* nuclear factor of activated T cells)
NLS 39, 42
NPC (*see* nuclear pore complex)
Npl3p 42
NRS (*see* nuclear retention-like sequence)
NTF2 86
nuclear
- bodies (Nbs) 32
- domains 10 (ND10) 32
- dots (Nds) 32
- export signal (NES) 2, 40–42, 47, 48, 83, 85, 86
- - dependent export 42
- factor of activated T cells (NFAT) 125, 126
- matrix 31, 37
- pore complex (NPC) 29, 30, 39, 47, 48, 84, 95
- - architecture 104, 105
- - functional status 105
- retention
- - intron-containing RNA 79, 80
- - MPMV RNA 80
- retention-like sequence (NRS) 39
nuclear tracks 32, 33
nucleo-cytoplasmic transport 95
nucleoporins 64, 84, 96
Nup 98 84, 86
NXT1/p15 86

O
ONYX-015 43
OSTR 126

P
p53 41, 43, 46
- proteolytic degradation 46
- status 43
p62 86
PHAX 89
PKI (*see* protein kinase inhibitor)
PML 44
- oncogenic domains (PODs) 32, 44
- - disruption 34
- - reorganization 34
- - structure 34, 44
poly(A)$^+$ RNA 42
polyadenylation 88
posttranscriptional gene regulation 77

protein kinase inhibitor (PKI) 125
protein phosphatase (PP2A) 35, 36

R
rAAV (adeno-associated virus vector) 41, 46
Rab 42
Ran 42, 44, 48
Ran-GAP 84
Ran-GDP 84, 86
Ran-GTP 30, 47, 48, 83, 84, 86
RCC1 44, 84
replicative foci 32, 33
retroviral transcription 78
retrovirus 77
– replication 78
Rev 29–31, 39, 40–42, 47, 48, 55, 61, 78–81
Rev NES 30
Rev response element (RRE) 30, 41, 56, 78, 80
Rev/Rex 41
Rev/RRE function 80, 81, 83, 84, 87
Rex 39, 41, 55, 61, 78
Rex response element (RxRE) 58, 78
ribonucleoprotein complexes (RNPs) 29, 37–39, 42, 48
Rip 42
Rip1p 42, 64
RIP/Rab 42, 64
RNA helicase 45, 85
RNPs (see ribonucleoprotein complexes)
RRE (see Rev response element)
RXL 46
RXL motif 41
RxRE (see Rex response element) 58, 78

S
5S rRNA 42
S phase 44, 46
Saccharomyces cerevisiae 42, 48
SAM (Src-associated in mitosis)-68 86
secondary structure, MPMV CTE 82

serine-arginine-rich 35
small nuclear ribonucleoprotein particles (snRNPs) 35
Sp100 44
SR proteins 36
Src 86
STAR proteins 88, 89
structural analysis 81
– MPMV CTE 81, 82
– SRV CTE 81

T
TIP-associated protein (TAP) 39, 42, 85, 86
TPL (*see* tripartite leader)
transcription inhibitors 120, 121
translation 44
transportin 85
tripartite leader (TPL) 27, 35, 37, 45
– assembly 35
– splicing 34
tubulin β-tubulin 36
type D simpler retroviruses 79, 80

U
U snRNAs 42
ubiquitin-dependent proteolytic degradation 46

V
VA RNAs 27
VAI RNAs 45
viral transcription/replication centers 48

X
Xenoput oocytes 82–84, 87

Y
yeast
– Gle1p 42
– Mex67p 42

Printing (Computer to Film): Saladruck, Berlin
Binding: Stürtz AG, Würzburg

Current Topics in Microbiology and Immunology

Volumes published since 1989 (and still available)

Vol. 216: **Rietschel, Ernst Th.; Wagner, Hermann (Eds.):** Pathology of Septic Shock. 1996. 34 figs. X, 321 pp. ISBN 3-540-61026-X

Vol. 217: **Jessberger, Rolf; Lieber, Michael R. (Eds.):** Molecular Analysis of DNA Rearrangements in the Immune System. 1996. 43 figs. IX, 224 pp. ISBN 3-540-61037-5

Vol. 218: **Berns, Kenneth I.; Giraud, Catherine (Eds.):** Adeno-Associated Virus (AAV) Vectors in Gene Therapy. 1996. 38 figs. IX,173 pp. ISBN 3-540-61076-6

Vol. 219: **Gross, Uwe (Ed.):** Toxoplasma gondii. 1996. 31 figs. XI, 274 pp. ISBN 3-540-61300-5

Vol. 220: **Rauscher, Frank J. III; Vogt, Peter K. (Eds.):** Chromosomal Translocations and Oncogenic Transcription Factors. 1997. 28 figs. XI, 166 pp. ISBN 3-540-61402-8

Vol. 221: **Kastan, Michael B. (Ed.):** Genetic Instability and Tumorigenesis. 1997. 12 figs.VII, 180 pp. ISBN 3-540-61518-0

Vol. 222: **Olding, Lars B. (Ed.):** Reproductive Immunology. 1997. 17 figs. XII, 219 pp. ISBN 3-540-61888-0

Vol. 223: **Tracy, S.; Chapman, N. M.; Mahy, B. W. J. (Eds.):** The Coxsackie B Viruses. 1997. 37 figs. VIII, 336 pp. ISBN 3-540-62390-6

Vol. 224: **Potter, Michael; Melchers, Fritz (Eds.):** C-Myc in B-Cell Neoplasia. 1997. 94 figs. XII, 291 pp. ISBN 3-540-62892-4

Vol. 225: **Vogt, Peter K.; Mahan, Michael J. (Eds.):** Bacterial Infection: Close Encounters at the Host Pathogen Interface. 1998. 15 figs. IX, 169 pp. ISBN 3-540-63260-3

Vol. 226: **Koprowski, Hilary; Weiner, David B. (Eds.):** DNA Vaccination/Genetic Vaccination. 1998. 31 figs. XVIII, 198 pp. ISBN 3-540-63392-8

Vol. 227: **Vogt, Peter K.; Reed, Steven I. (Eds.):** Cyclin Dependent Kinase (CDK) Inhibitors. 1998. 15 figs. XII, 169 pp. ISBN 3-540-63429-0

Vol. 228: **Pawson, Anthony I. (Ed.):** Protein Modules in Signal Transduction. 1998. 42 figs. IX, 368 pp. ISBN 3-540-63396-0

Vol. 229: **Kelsoe, Garnett; Flajnik, Martin (Eds.):** Somatic Diversification of Immune Responses. 1998. 38 figs. IX, 221 pp. ISBN 3-540-63608-0

Vol. 230: **Kärre, Klas; Colonna, Marco (Eds.):** Specificity, Function, and Development of NK Cells. 1998. 22 figs. IX, 248 pp. ISBN 3-540-63941-1

Vol. 231: **Holzmann, Bernhard; Wagner, Hermann (Eds.):** Leukocyte Integrins in the Immune System and Malignant Disease. 1998. 40 figs. XIII, 189 pp. ISBN 3-540-63609-9

Vol. 232: **Whitton, J. Lindsay (Ed.):** Antigen Presentation. 1998. 11 figs. IX, 244 pp. ISBN 3-540-63813-X

Vol. 233/I: **Tyler, Kenneth L.; Oldstone, Michael B. A. (Eds.):** Reoviruses I. 1998. 29 figs. XVIII, 223 pp. ISBN 3-540-63946-2

Vol. 233/II: **Tyler, Kenneth L.; Oldstone, Michael B. A. (Eds.):** Reoviruses II. 1998. 45 figs. XVI, 187 pp. ISBN 3-540-63947-0

Vol. 234: **Frankel, Arthur E. (Ed.):** Clinical Applications of Immunotoxins. 1999. 16 figs. IX, 122 pp. ISBN 3-540-64097-5

Vol. 235: **Klenk, Hans-Dieter (Ed.):** Marburg and Ebola Viruses. 1999. 34 figs. XI, 225 pp. ISBN 3-540-64729-5

Vol. 236: **Kraehenbuhl, Jean-Pierre; Neutra, Marian R. (Eds.):** Defense of Mucosal Surfaces: Pathogenesis, Immunity and Vaccines. 1999. 30 figs. IX, 296 pp. ISBN 3-540-64730-9

Vol. 237: **Claesson-Welsh, Lena (Ed.):** Vascular Growth Factors and Angiogenesis. 1999. 36 figs. X, 189 pp. ISBN 3-540-64731-7

Vol. 238: **Coffman, Robert L.; Romagnani, Sergio (Eds.):** Redirection of Th1 and Th2 Responses. 1999. 6 figs. IX, 148 pp. ISBN 3-540-65048-2

Vol. 239: **Vogt, Peter K.; Jackson, Andrew O. (Eds.):** Satellites and Defective Viral RNAs. 1999. 39 figs. XVI, 179 pp. ISBN 3-540-65049-0

Vol. 240: **Hammond, John; McGarvey, Peter; Yusibov, Vidadi (Eds.):** Plant Biotechnology. 1999. 12 figs. XII, 196 pp. ISBN 3-540-65104-7

Vol. 241: **Westblom, Tore U.; Czinn, Steven J.; Nedrud, John G. (Eds.):** Gastroduodenal Disease and Helicobacter pylori. 1999. 35 figs. XI, 313 pp. ISBN 3-540-65084-9

Vol. 242: **Hagedorn, Curt H.; Rice, Charles M. (Eds.):** The Hepatitis C Viruses. 2000. 47 figs. IX, 379 pp. ISBN 3-540-65358-9

Vol. 243: **Famulok, Michael; Winnacker, Ernst-L.; Wong, Chi-Huey (Eds.):** Combinatorial Chemistry in Biology. 1999. 48 figs. IX, 189 pp. ISBN 3-540-65704-5

Vol. 244: **Daëron, Marc; Vivier, Eric (Eds.):** Immunoreceptor Tyrosine-Based Inhibition Motifs. 1999. 20 figs. VIII, 179 pp. ISBN 3-540-65789-4

Vol. 245/I: **Justement, Louis B.; Siminovitch, Katherine A. (Eds.):** Signal Transduction and the Coordination of B Lymphocyte Development and Function I. 2000. 22 figs. XVI, 274 pp. ISBN 3-540-66002-X

Vol. 245/II: **Justement, Louis B.; Siminovitch, Katherine A. (Eds.):** Signal Transduction on the Coordination of B Lymphocyte Development and Function II. 2000. 13 figs. XV, 172 pp. ISBN 3-540-66003-8

Vol. 246: **Melchers, Fritz; Potter, Michael (Eds.):** Mechanisms of B Cell Neoplasia 1998. 1999. 111 figs. XXIX, 415 pp. ISBN 3-540-65759-2

Vol. 247: **Wagner, Hermann (Ed.):** Immunobiology of Bacterial CpG-DNA. 2000. 34 figs. IX, 246 pp. ISBN 3-540-66400-9

Vol. 248: **du Pasquier, Louis; Litman, Gary W. (Eds.):** Origin and Evolution of the Vertebrate Immune System. 2000. 81 figs. IX, 324 pp. ISBN 3-540-66414-9

Vol. 249: **Jones, Peter A.; Vogt, Peter K. (Eds.):** DNA Methylation and Cancer. 2000. 16 figs. IX, 169 pp. ISBN 3-540-66608-7

Vol. 250: **Aktories, Klaus; Wilkins, Tracy, D. (Eds.):** Clostridium difficile. 2000. 20 figs. IX, 143 pp. ISBN 3-540-67291-5

Vol. 251: **Melchers, Fritz (Ed.):** Lymphoid Organogenesis. 2000. 62 figs. XII, 215 pp. ISBN 3-540-67569-8

Vol. 252: **Potter, Michael; Melchers, Fritz (Eds.):** B1 Lymphocytes in B Cell Neoplasia. 2000. XIII, 326 pp. ISBN 3-540-67567-1

Vol. 253: **Gosztonyi, Georg (Ed.):** The Mechanisms of Neuronal Damage in Virus Infections of the Nervous System. 2001. approx. XVI, 270 pp. ISBN 3-540-67617-1

Vol. 254: **Privalsky, Martin L. (Ed.):** Transcriptional Corepressors. 2001. 25 figs. XIV, 190 pp. ISBN 3-540-67569-8

Vol. 255: **Hirai, Kanji (Ed.):** Marek's Disease. 2001. 22 figs. XII, 294 pp. ISBN 3-540-67798-4

Vol. 256: **Schmaljohn, connie S.; Nichol, Stuart T. (Eds.):** Hantaviruses. 2001, 24 figs. XI, 196 pp. ISBN 3-540-41045-7

Vol. 257: **van der Goot, Gisou (Ed.):** Pore-Forming Toxins, 2001. 19 figs. IX, 166 pp. ISBN 3-540-41386-3

Vol. 258: **Takada, Kenzo (Ed.):** Epstein-Barr Virus and Human Cancer. 2001. 38 figs. XII, 222 pp. ISBN 3-540-41506-8